INTERNET OF THINGS FOR AGRICULTURE 4.0

Impact and Challenges

INTERNET OF THINGS FOR AGRICULTURE 4.0

Impact and Challenges

Edited by
Rajesh Singh, PhD
Amit Kumar Thakur, PhD
Anita Gehlot, PhD
Ajay Kumar Kaviti, PhD

A·AP | APPLE
ACADEMIC
PRESS

First edition published 2022

Apple Academic Press Inc.
1265 Goldenrod Circle, NE,
Palm Bay, FL 32905 USA

4164 Lakeshore Road, Burlington,
ON, L7L 1A4 Canada

CRC Press
6000 Broken Sound Parkway NW,
Suite 300, Boca Raton, FL 33487-2742 USA

2 Park Square, Milton Park,
Abingdon, Oxon, OX14 4RN UK

Library and Archives Canada Cataloguing in Publication

Title: Internet of things for agriculture 4.0 : impact and challenges / edited by Rajesh Singh, PhD, Amit Kumar Thakur, PhD, Anita Gehlot, PhD, Ajay Kumar Kaviti, PhD.
Names: Singh, Rajesh (Electrical engineer), editor. | Kumar Thakur, Amit, editor. | Gehlot, Anita, editor. | Kaviti, Ajay Kumar, editor.
Description: First edition. | Includes bibliographical references and index.
Identifiers: Canadiana (print) 20210319569 | Canadiana (ebook) 2021031964X | ISBN 9781774630020 (hardcover) | ISBN 9781774639214 (softcover) | ISBN 9781003161097 (ebook)
Subjects: LCSH: Agricultural informatics. | LCSH: Internet of things—Industrial applications. | LCSH: Wireless LANs—Industrial applications.
Classification: LCC S494.5.D3 I58 2022 | DDC 630.2/0854678—dc23

Library of Congress Cataloging-in-Publication Data

Names: Singh, Rajesh, 1980- editor. | Thakur, Amit Kumar, editor. | Gehlot, Anita, editor. | Kaviti, Ajay Kumar, 1977- editor.
Title: Internet of things for agriculture 4.0 : impact and challenges / Rajesh Singh, Amit Kumar Thakur, Anita Gehlot, Ajay Kumar Kaviti.
Description: First edition. | Palm Bay, FL, USA : Apple Academic Press, 2022. | Includes bibliographical references and index. | Summary: "This new book provides an insightful look at the varied and exciting uses and applications of Wi-Fi and the Internet of Things in agriculture. With internet-enabled communications becoming more widely available, farms and agricultural establishments can take advantage of these new technologies for a wide range of farm operations, such as crop management, farm vehicle tracking, livestock monitoring, storage monitoring, and more. The collected data from these devices can be stored in the cloud system or server and accessed by the farmers via the internet or mobile phones. This book shows the many benefits to farmers from applying IoT, including better utilizing information for monitoring crops, optimizing water use, planning effective fertilization strategies, and saving time and reducing the operation expenses. Topics include using IoT for vertical farming, IoT-based smart irrigation system, landslide susceptibility assessment, automated aeroponics systems, crop survival analysis, and more. The volume also considers the challenges of IoT in agriculture, such as the requirements of applications of wireless sensor networks, the threat of attacks and the detection of vulnerabilities in wireless sensor networks, and more. Internet of Things for Agriculture 4.0: Impact and Challenges provides a better understanding of the time- and resourcing-saving benefits of wireless sensors and remote monitoring devices in agriculture. The volume will be useful for those involved in agricultural operations as well as scientists and researchers, and faculty and students in agriculture and computer and information science engineering"-- Provided by publisher.
Identifiers: LCCN 2021045081 (print) | LCCN 2021045082 (ebook) | ISBN 9781774630020 (hardback) | ISBN 9781774639214 (paperback) | ISBN 9781003161097 (ebook)
Subjects: LCSH: Internet of things. | Agricultural innovations. | Agriculture--Automation.
Classification: LCC S494.5.D3 I586 2022 (print) | LCC S494.5.D3 (ebook) | DDC 338.1--dc23
LC record available at https://lccn.loc.gov/2021045081
LC ebook record available at https://lccn.loc.gov/2021045082

ISBN: 978-1-77463-002-0 (hbk)
ISBN: 978-1-77463-921-4 (pbk)
ISBN: 978-1-00316-109-7 (ebk)

About the Editors

Rajesh Singh, PhD, is currently associated with Lovely Professional University as Professor with more than 16 years of experience in academics. He has been awarded as gold medalist in MTech from RGPV, Bhopal (M.P) India and honors in his B.E. from Dr. B. R. Ambedkar University, Agra (U.P), India. His area of expertise includes embedded systems, robotics, wireless sensor networks, and Internet of Things. He has been honored as a keynote speaker and session chair at international/national conferences, faculty development programs, and workshops. He has 152 patents in his account. He has published more than 100 research papers in refereed journals/conferences and 24 books in the area of embedded systems and Internet of Things with reputed publishers. He is editor of a special issues published by the AISC book series, Springer in 2017 & 2018, and IGI Global in 2019.

Under his mentorship, students have participated in national/international competitions, including "Innovative Design Challenge Competition" by Texas and DST and Laureate Award of Excellence in Robotics Engineering, Madrid, Spain, in 2014 and 2015. His team has been the winner of "Smart India Hackathon-2019" hardware version conducted by MHRD, Government of India, for the problem statement of Mahindra & Mahindra. Under his mentorship, is student team got the "InSc award 2019" under the students projects program. He has been awarded with a "Gandhian Young Technological Innovation (GYTI) Award," as mentor to "On Board Diagnostic Data Analysis System-OBDAS," appreciated under "Cutting Edge Innovation" during the Festival of Innovation and Entrepreneurship at Rashtrapati Bhavan, India, in 2018. He has been honored with a "Certificate of Excellence" from 3rd faculty branding awards-15, organized by EET CRS research wing for excellence in professional education and industry, for the category "Award for Excellence in Research" (2015) and young investigator award at the International Conference on Science and Information in 2012.

Amit Kumar Thakur, PhD is currently associated with Lovely Professional University as Associate Professor with more than 17 years of experience in academics and research. His area of expertise includes renewable energy technologies, thermal energy, robotics, and Internet of Things. He has organized and conducted a number of workshops, summer internships, and expert lectures for students as well as faculty. He has published around 24 research papers in refereed journals/conferences. He is a life member of The Indian Society for Technical Education (ISTE), faculty member of the Society of Automotive Engineers (SAE), life member of The Aeronautical Society of India (AMeASI). He is an editorial board member and reviewer for various reputed journals. He was the research sub-coordinator for ADRDE, DRDO in the project of design and development of the deceleration system for a space capsule recovery experiment (SRE). He has five patents to his name. He is the editor of two books, one from CRC Press for (Taylor & Francis) and one from Nova Science Publishers.

Anita Gehlot, PhD, is currently associated with Lovely Professional University as Associate Professor with more than 12 years of experience in academics. Her area of expertise includes embedded systems, wireless sensor networks, and Internet of Things. She has been honored as a keynote speaker and session chair at international/national conferences, faculty development programs, and workshops. She has 132 patents in her account. She has published more than 70 research papers in refereed journals/conferences and 24 books in the area of embedded systems and Internet of Things with reputed publishers including CRC/Taylor & Francis, Narosa, GBS, IRP, NIPA, River Publishers, Bentham Science, and RI publication.

She is editor of a special issue published by AISC book series, Springer in 2018, and IGI Global in 2019. She has been awarded with a "certificate of appreciation" from the University of Petroleum and Energy Studies for exemplary work. Under her mentorship, her student team got "InSc award 2019" under the students projects program. She has been awarded with a "Gandhian Young Technological Innovation (GYTI) Award," as mentor to "On Board Diagnostic Data Analysis System-OBDAS," appreciated under "Cutting Edge Innovation" during the Festival of Innovation and Entrepreneurship at Rashtrapati Bhavan, India, in 2018.

Ajay Kumar Kaviti, PhD, is an Associate Professor in the Department of Mechanical Engineering, VNR Vignana Jyothi Institute of Engineering & Technology, Hyderabad, India. He did his MTech in 2003 and PhD in 2010 at Maulana Azad National Institute of Technology, Bhopal. He has more than 16 years of experience in teaching and research. He has multidisciplinary research interests in the areas of renewable energy, alternative fuels, and composites. He has guided 15 MTech projects and is currently guiding PhD students. He has 40 publications to his credit in highly reputed international journals along with 20 publications in proceedings of national and international conferences. He is on review board of many prestigious national and international journals and editorial board member for various international journals.

Contents

Contributors

Swapnil Bagwari
School of Electronics and Electrical Engineering, Lovely Professional University, Phagwara, Punjab, India

Anita Gehlot
School of Electronics and Electrical Engineering, Lovely Professional University, Phagwara, Punjab, India

Himani Jerath
School of Electronics & Electrical Engineering, Lovely Professional University, Phagwara, Punjab, India

G. S. Mahesh
Department of Electrical and Electronics Engineering, Sree Vidyanikethan Engineering College, Tirupati, Andhra Pradesh, India

Yogesh Pant
Department of Computer and Information Studies, Himalayan Institute of Science and Technology, SRHU, Dehradun, Uttarakhand, India

P. Raja
School of Electronics & Electrical Engineering, Lovely Professional University, Phagwara, Punjab, India

B. Vidheya Raju
Department of Electronics and Communication Engineering, Aditya Engineering College (A), Surampalem, Andhra Pradesh, India

Geeta Rana
Swami Rami Himalayan University, Dehradun, Uttarakhand, India

P. S. Ranjit
Department of Mechanical Engineering, Aditya Engineering College(A), Surampalem, Andhra Pradesh, India

Ch Sai Dinesh Reddy
Lovely Professional University, Phagwara, Jalandhar 144411, India

Mohit Kumar Saini
Department of Computer Science, Doon Business School, Dehradun, Uttarakhand, India

Rakesh Kumar Saini
School of Computing, DIT University, Dehradun, Uttarakhand, India

Bharti Sharma
School of Computing, DIT University, Dehradun, Uttarakhand, India

Ravindra Sharma
Swami Rama Himalayan University, Dehradun, Uttarakhand, India

Dushyant Kumar Singh

School of Electronics & Electrical Engineering, Lovely Professional University, Phagwara, Punjab, India

Rajesh Singh

School of Electronics and Electrical Engineering, Lovely Professional University, Phagwara, Punjab, India

M. Suresh

Lovely Professional University, Phagwara, Punjab, India

Mahendra Swain

Department of Electronics and Communication Engineering, Lovely Professional University, Phagwara, Punjab, India

Abbreviations

ADCs	analog-to-digital converters
ADR	adaptive data rate
AI	artificial intelligence
AIDE	Arduino Integrated Development Environment
API	application program interface
ASICS	application specific circuits
AT	attention
AUC	area under curve
EEPROM	electrically erasable programmable read only memory
FAO	Food and Agriculture Organization
FHSS	frequency hop spread spectrum
FN	false negative
FP	false positive
GNSS	global navigation satellite system
GPIO	general input/output
GPS	global positioning system
GSM	Global Mobile Communications Service
HVAC	heating, ventilation and air conditioning
IoT	Internet of Things
ISM	industrial, scientific and medical
LMD	labor monitoring device
LR-WPANs	low-rate wireless personal area networks
MAC	medium access
MANETs	mobile ad-hoc networks
MIPS	million instructions per second
MSE	mean square error
NC	no connection
NTC	negative temperature coefficient
OSI	open systems interconnection
PANs	personal area networks
PCB	printed circuit board
PHY	physical radio
RAM	random access memory

RFID	radio frequency ID
ROM	read only memory
SA	smart agriculture
SCL	serial clock pin
SDA	serial data pin
SPI	serial peripheral interface
SVM	support vector machine
TN	true negative
TP	true positive
TTN	the things network
VF	vertical farming
WSN	wireless sensor network
WWW	World Wide Web

Foreword

Internet of Things (IoT) and aerial mapping are nowadays being used very much in agriculture. IoT plays a crucial role in smart agriculture. Smart farming is an emerging concept, because IoT sensors are capable of providing information about their agriculture fields. This book focuses on carrying out various works of researchers on making use of evolving technology, i.e., IoT and smart agriculture using automation.

Dr. Stanislaw Szwaja,
Associate Professor
Czestochowa University of Technology
phone: (+48) 885840483
Email: szwaja@imc.pcz.czest.pl

Automation of farm activities can transform the agricultural domain from being manual and static to intelligent and dynamic leading, to higher production with lesser human supervision. This book describes the benefit of the Internet of Things and the smart technologies. In addition, it addresses the impacts and challenges of Internet of Things in Agriculture 4.0.

Dr. Roopesh Mehra
Project Manager R&D
Science Research Institute
Weichai Power Company Limited
Weifang, China
Email: roopeshmehra.ind@gmail.com

Preface

This book aims to provide a better understanding of the opportunities and resource savings of using wireless sensors and the remote use of monitoring devices on farms. The Internet of Things (IoT), or Internet-enabled communications, was simply explained by Jacob Morgan, who stated that broadband Internet is widely available and more devices now come with Wi-Fi capabilities and sensors with them.

The devices can create a "storm" around a common Wi-Fi. In agriculture, IoT applications include farm vehicle tracking, livestock monitoring, storage monitoring, and other farm operations. The collected data can be stored in the cloud system or on a server and accessed by farmers via the Internet or their mobile phones. IoT has been applied in agriculture in farming, in fisheries and aquaculture, in animal food consumption, in agri-food supply chain, in greenhouse horticulture, and in livestock farming.

The book comprises total 11 chapters.

Editors are thankful to all the contributors and to the publisher for their support.

— The Editors

CHAPTER 1

Vertical Farming Trends and Challenges: A New Age of Agriculture Using IoT and Machine Learning

MAHENDRA SWAIN*

Department of Electronics and Communication Engineering, Lovely Professional University, Phagwara, Punjab, India

*E-mail: er.mahendraswain@gmail.com

ABSTRACT

Vertical farming (VF) is a new age of agriculture technique. It has the potential to fulfill the food requirement in the future. Looking toward the current trend of agriculture, it seems to be that VF could enable all the vertices of farming in various dimensions. It is an unconventional way of agriculture to meet the food requirement as the farming lands are getting shirked day by day. Implementing cutting-edge technologies like IoT (Internet of things), AI (artificial intelligence), and machine learning, the productivity and quality factors would enhance VF. This chapter illustrates how advanced technology is integrated into the farming for overall growth of farmer and economy of a country. By deploying sensor nodes, farming monitoring is reliable and effective way to handle day-to-day activities in farming land.

1.1 INTRODUCTION

The estimation tells that by 2050 there will be 9 billion population across the world. Most of the population is moving toward urban areas. As per study, in 2018, 800 million ha of land is based on soil-based agriculture.

Gradually, the farming land is decreasing due to civilization of modern world.[1] As the population increases, the food demands will also increase. Whereas we do not have enough farming land to meet the food requirement, so this is the time to switch to unconventional way of agriculture, that is, vertical farming (VF), hydroponic, and aeroponic.[2] Adapting this kind of farming technology with the help of cutting-edge technologies like Internet of things (IoT) and artificial intelligence (AI) could be the revolutionary approach. VF is more reliable and suitable because it does not require soil and sunlight. It grows in vertical layers stacked one upon another. The closed ecosystem allows us to control all the activities inside this. Moreover, it can yield maximum from minimum area, more quality concerned.[3-5] The most advantage is it does get affected by environmental conditions and natural calamities. The uses of IoT would enable monitoring of the farms from a remote distance. A fully autonomous ecosystem will help the plant to grow precisely and full of nutrients. Although it does have so many advantages, there are various challenges in VF given in the following sections.

1.1.1   CHALLENGES IN VERTICAL FARMING

- Limited number of crops can be cultivated using VF
- Slow-growing plants like rice, paddy cannot be cultivated
- Deep-rooted plants like potato are not suitable
- High space–requiring plants such as corn is avoided in VF
- Lighting is another issue in this for photosynthesis purpose; plants require light with suitable amount of intensity (lux: unit of light intensity measurement)
- Efficient design architecture of VF
- Monitoring water and nutrients required for the plants
- Controlling HVAC (heating, ventilation, and air conditioning) system in closed ecosystem
- Recycling the waste

The abovementioned challenges are faced in VF.[6] To overcome these challenges, it is necessary to adapt new technologies like automatic HVAC ecosystem using IoT and machine learning to select the crops that are suitable according to the nutrients requirement to grow healthy plants.[7]

1.2 VERTICAL FARMING ECOSYSTEM

FIGURE 1.1 Examples of vertical farming.
Source: a) Photo by Valcenteu. https://creativecommons.org/licenses/by-sa/3.0/ b) Photo by Satoshi KINOKUNI - https://www.flickr.com/photos/nikunoki/38459770052/ https://creativecommons.org/licenses/by/2.0/ c) Photo by Benjamin D. Esham / Wikimedia Commons. https://creativecommons.org/licenses/by-sa/3.0/us/deed.en

TABLE 1.1 Various Vertical Farming Ecosystem across Globe.

Sl. no.	Name	Location	Agriproducts	Size	Technology used
1	The Plant Vertical Farm	Chicago	Mushroom, bakery	1 Lakh ft²	Aquaponics Hydroponic Natural sun source Recycling waste into energy
2	Sky Greens Farms	Singapore	Leafy green vegetables	600 m	Aeroponic system Low-carbon hydraulic driven

TABLE 1.1   *(Continued)*

Sl. no.	Name	Location	Agriproducts	Size	Technology used
3	VertiCrop	Canada	Leafy greens and strawberries	16 ac	Fully automated with closed loop HVAC system Natural and artificial light source
4	Republic of South Korea VF	South Korea	Leafy green vegetables, almost wheat	450 m²	Renewable resources like solar LED light source
5	Nuvege Plant Factory	Japan	Leafy greens vegetables	30,000 × 57,000 ft²	Automated rack LED light source Hydroponics
6	AeroFarms	New Jersey	250 types of herbs such as kale and mizuna	20,000 ft² with 35 rows and 12 levels	LED light source Sensor tracking system for growing plant Recycle water techniques

Figure 1.1 shows the vertical farming plant. All the three images are the type of VF plants. There is also worldwide popularity of hydroponic farming.[8] In Table 1.1, details of VF worldwide are given which shows the popularity of VF. It is different from traditional agriculture techniques.[9]

The proposed system comprises the subimmerse sink node, LoRa gateway, cloud server as shown in Figure 1.2. Different environmental parameters like temperature, humidity, PH level, and fresh water are sent to the cloud server through LoRa communication module.[10,11] The subimmerse sink node transmits the data to the gateway node. The gateway node logs the data into the cloud server via Wi-Fi and also logs the data to local server via LoRa.[12] With the assistance cloud server, the user can access the data of hydroponic farming through mobile and Web.[13] The real-time data of hydroponic system for every successful time interval is logged into cloud server and it will help the user to analyze the environmental parameter of farming system in a systematic manner. After analyzing sensor data, corrective action could be taken like the addition of nutrient to the water solution and maintaining the level of N_2, O_2 in liquid solution.

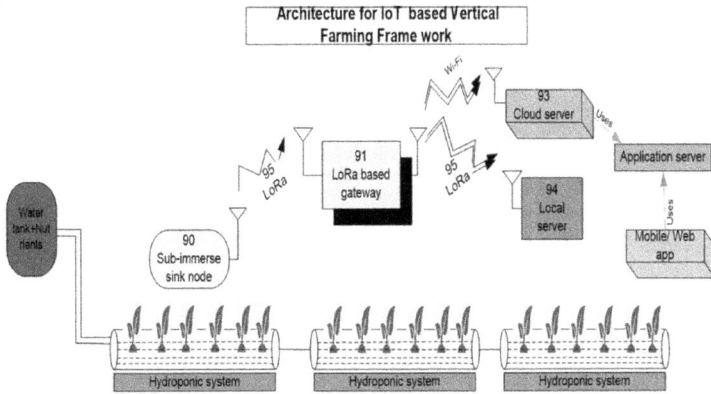

FIGURE 1.2   System for real-time monitoring of hydroponic farming system.

1.3   INTEGRATION OF LORAWAN WITH VF

ADR (adaptive data rate) system and distinctive (spreading factor) SF's control appropriation will improve arrange the adaptability in LoRaWAN.[14] A sensor node detail is shown in Figure 1.3. The localization principle of sensors organize versatility bottleneck is utilization of Aloha based MAC convention. What is more, the half-duplex activity of passages just as obligation cycle restriction expanded the negative effect on system adaptability.

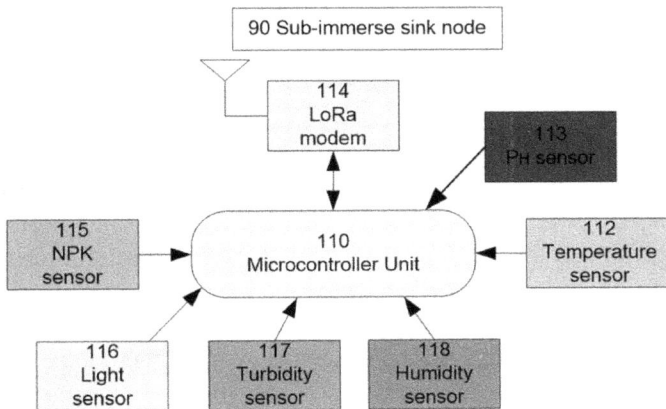

FIGURE 1.3   Sensor node.

In LoRa architecture, nodes are interconnected with star-to-star topology. It supports three various endpoint classes. These are listed below.

(a) Class A
(b) Class B
(c) Class C

Class A: It is a bidirectional device connected at end node. Both transmission and reception are possible in this class. In order to do this, end devices are provided with uplink and downlink windows.[15] One is for transmission and another is for reception. Transfer occurs through some dedicated time slots assigned by the user. Uplink transmission helps send the data from end devices to the server and downlink waits until the data get received.[16] This consumes very less power to enhance battery efficiency.

Class B: It is superior to Class A end device. It does have bidirectional end devices along with a receiver slot, which follows scheduled mechanism. It has an extra receive window that looks after time slots allocated at uplink window. Proper synchronization is maintained between gateway and end devices to ensure that the gateway listens to the end devices, that is, all data are received at the gateway.

Class C: It does have the features of Class A and with maximum receiving slots. This class of LoRa network supports reception window with regular reception of data from end devices. Class C is preferred when large amount of data needs to be received at gateway. Class C closes its window during transmission only.

1.3.1 MESSAGE PASSING OVER LORA

Message passing over LoRaWAN is very crucial, when end devices pass message to gateway in coverage network area. It receives the message and acknowledgment signal passed to the devices. To design cost-effective network, adapting localization algorithm is used. A minimum number of devices can be deployed to collect the data. Gateways approve the received data that is coming from end devices. This way error can be reduced in the LoRa network.[17] Several layers of encryption have been done to keep the data secure. These are listed below:

• Unique Network key (EUI64): It is responsible to keep the data safe and secure when data passes through various networks

- Unique Application key (EUI64): This key takes care of the data when data transmits from one node to another in the application layer of LoRa network
- Specific key for end device (EUI128): Device dedicated keys make authentication of devices connected to LoRa network wirelessly. Each device allocated with a specific key.

Working:

Transmission symbol rate in LoRa network calculated using the following formula.

$$\text{Symbol rate}\,(Rs) = \text{SF} \times \frac{\text{BW}}{2^{\text{SF}}}$$

where SF is the spreading factor in LoRa network and BW is the bandwidth in hertz.

1.4 SIMULATION OF VF WITH LORAWAN

Data transmission occurs in the LoRa-following data frame structure mentioned in Figure 1.4. Preamble is the program code that needs to be transferred.[18] PHY header mechanism applied to the data packets. CRC is the cyclic redundancy check to take care of the same correct data bits received. Payload contains LoRaWAN or MAC data bits.

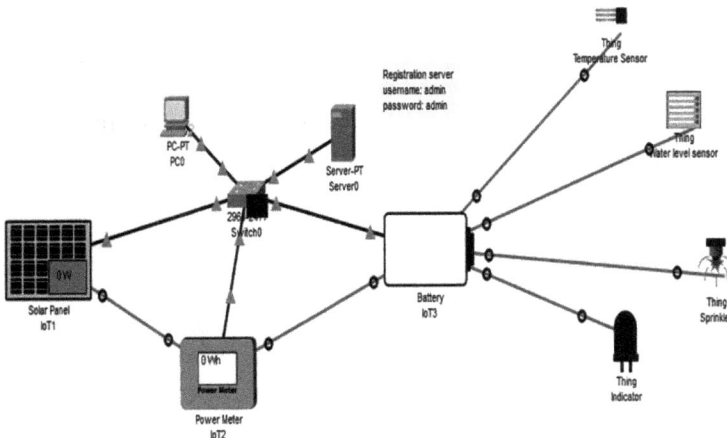

FIGURE 1.4   Simulated IoT-enabled network with sensor interfacing and energy harvesting.

There are various LoRa simulators that can be used to simulate the data. Table 1.1 shows simulators supporting different platforms and code.

Simulator features have been mentioned in Figure 1.5 with respect to physical model designed in the simulator.

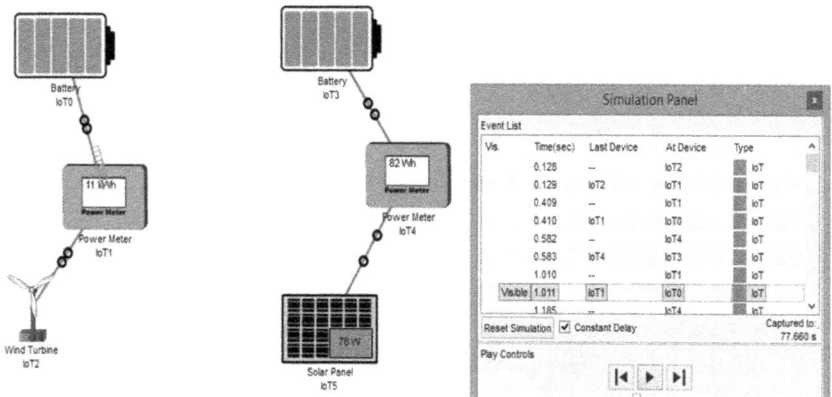

FIGURE 1.5   Renewable energy harvesting techniques in vertical farming from solar and wind.

The block diagram of LoRa transmitter has been shown in Figure 1.7. It comprises power supply unit both 5 and 12 V variants. ATMEGA328 controller[19] has been used in this transmitter and custom fabricated on PCB (printed circuit board). There are multiple pins dedicated for sensor interfacing both analog and digital sensors such as pH, ultrasonic, temperature, humidity, pressure sensor, rainfall sensor, and gas sensors. Controller has been interfaced with LoRa module. The data coming from sensors gets transmitted to the nearest local gateway (Fig. 1.6).

1.4.1   END DEVICE

The block diagram consists of power supply unit. On-board power supply pins +5 Volt, +12 Volts. ATMEGA328 has been interfaced with liquid-crystal display where all data can be visualized. LoRa module has been interfaced to transmit the data.

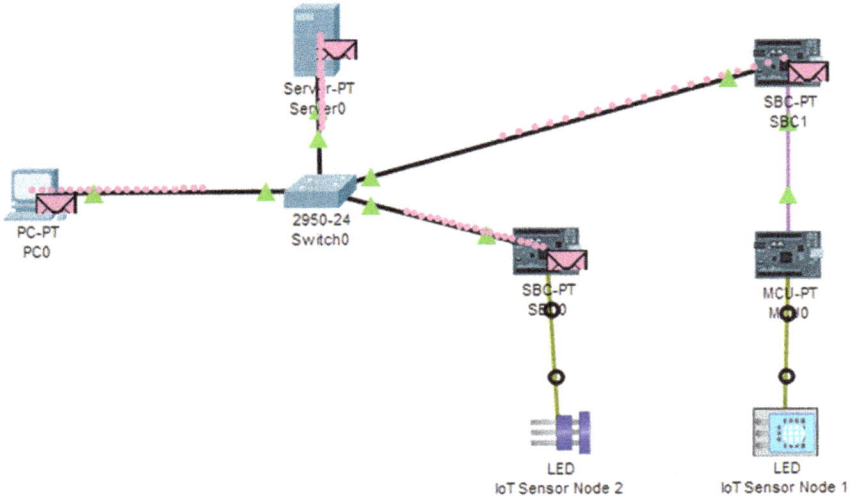

FIGURE 1.6   End nodes interfaced with sensor modules and gateway.

1.5   HARDWARE PLATFORM

A customized LoRa test bed has been designed to validate test code. Figure 1.8 shows LoRa customized transmitter module. It has dedicated analog and digital pins to interface various sensors to it. It is treated as end device in reference to LoRa architecture. It is operated on rechargeable battery and can be deployed into the agriculture field for field monitoring. The LoRa module has been fabricated with ATMEGA328P. PCB design for the same has been done PCB trace software (Fig. 1.7).

FIGURE 1.7   Customized LoRa end device node.

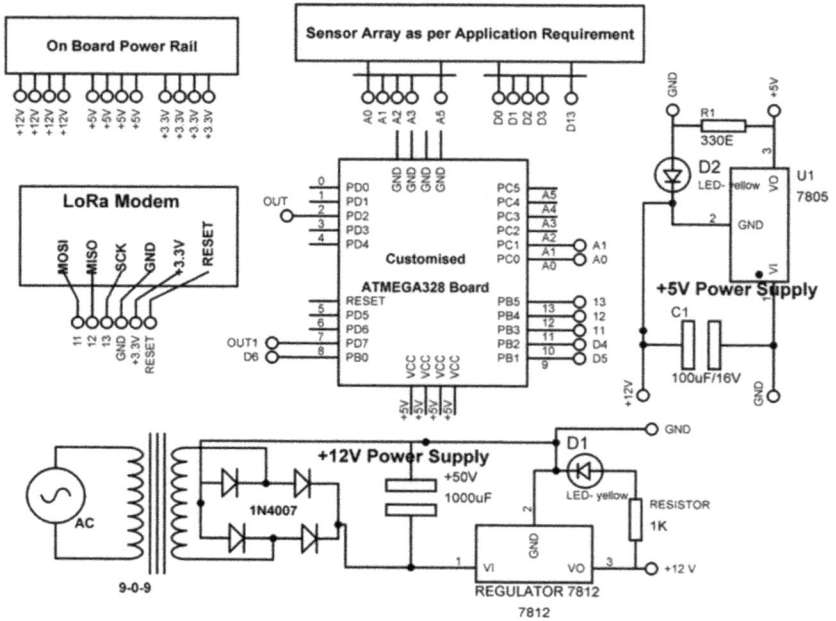

FIGURE 1.8   Circuit for LoRa transmitter node.

The circuit diagram shown in Figure 1.8 is for LoRa transmitter. The detail circuit diagram has been elaborated next. It comprises ATMEGA328 microcontroller, power supply unit, LoRa modem, sensor array strip for interfacing various sensors.

ATMEGA328 is a high-end 8-bit RISC microcontroller having 32 kb of ISP flash memory with read and write capability, 23 general-purpose input–output lines, and 10-bit ADC. Controller operates voltage between 1.8 and 5 V. Power supply unit consists of a 9-0-9 AC step-down transformer along with bridge rectifier that consists of four IN4007 diodes. The output voltage then passed through the regulated with regulated ICs (LM7812, LM7805). LM7812 is a very low-cost positive-voltage regulator IC with 12-V output. It has three terminals, IN, COM and OUT. IN is for input, COM is for ground, and OUT is to get the output from IC. An indicator has been connected to check the power supply unit working status, whereas LM7805 is a positive-voltage regulator IC with 5-V output voltage. Both of these IC belong to the same 78xx IC family.

LoRa modem SX1278 is a transceiver manufactured by SEMTECH with different operating frequencies from 137 to 525 MHz with low power consumption. The features of LoRa modem has been mentioned below:

- 168 dB link budget in LoRa
- +20 dB m to 100 mW constant RF output versus V supply
- +14 dB m high-efficiency PA
- Transmission bit rate 300 kbps
- Immunity toward noise interference
- Low receiver current of 9.9 mA, 200 nA retention in register
- FSK, GFSK, MSK, GMSK, LoRa, and OOK modulation
- Built-in bit synchronizer for clock recovery
- Automatic RF sense and CAD with ultrafast AFC
- Data packet rate up to 256 bytes with CRC
- Low-battery indicator and temperature sensor is inbuilt on modem

LoRa modem has been interfaced to ATMEGA328 with MOSI, MISO, SCLK and ground. Sensor array has multiple analog and digital pins to interface with digital and analog sensors such as DHT11, rainfall sensor, soil pH sensor, soil humidity sensor, pressure sensor, and gas sensors.

In Figure 1.9, both transmitter and receiver modules have been shown. Receiver node has LoRa module interface with 16 × 4 LCD to visualize sensor data from transmitting node. There are eight control buttons, to control various actuators connected to transmitter node.

FIGURE 1.9 Customized LoRa transmitter and receiver with LCD.

1.6   ANALYSIS USING MACHINE LEARNING

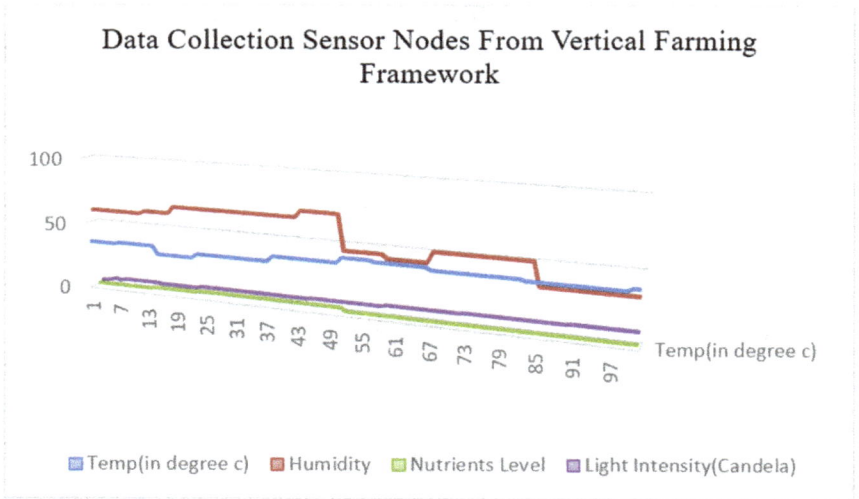

FIGURE 1.10   Data collection from sensor nodes in closed ecosystem of VF.

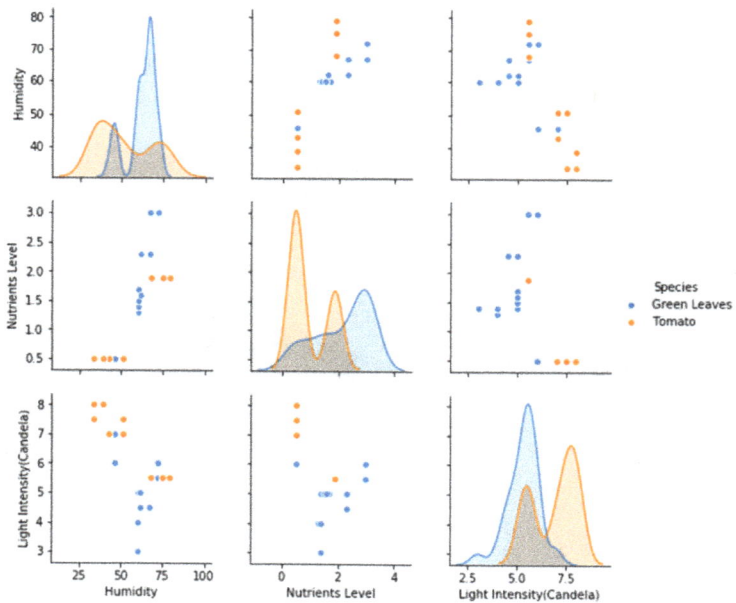

FIGURE 1.11   Data analysis using supervised learning algorithm for prediction.

The hyperplane separates data into two clusters by eliminating data overfitting and underfitting, which has been shown in Figure 1.11. If the data is nonlinear in nature, SVM uses kernel trick where data gets divided into three-dimensional view. Kernel tricking help the algorithm to identify the right hyperplane with minimize error. Further, a confusion matric is calculated to calculate the mean square error with respect to the training data and testing data.

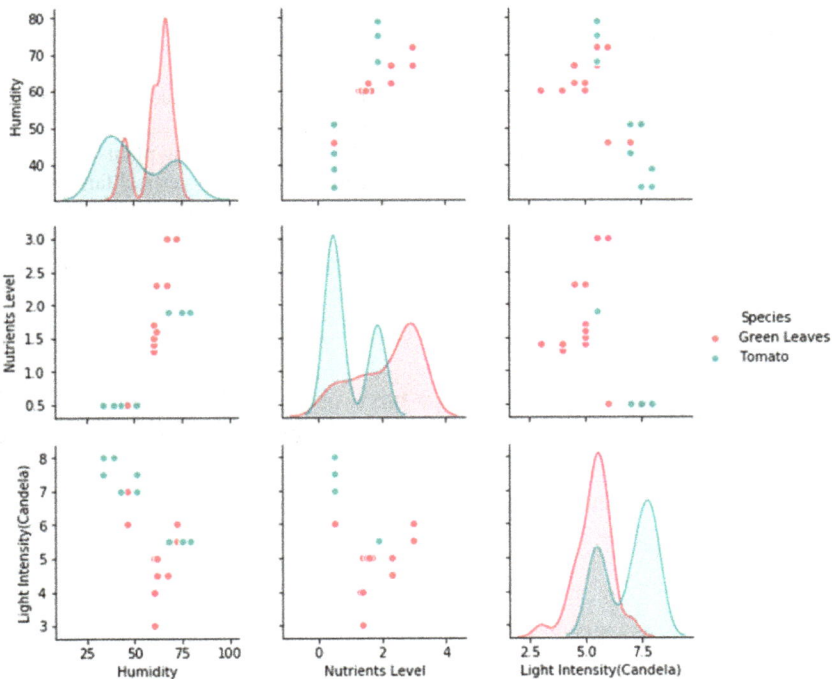

FIGURE 1.12   Data analysis using supervised learning algorithm for prediction.

Data mining helps to choose the appropriate parameter which is more relevant to the output of the farming. As shown in Figure 1.12, the parameters like temperature, humidity, nutrients level, and light intensities are playing major role in the production of green leafs and tomatoes in VF. Monitoring such parameters is very crucial. As per the data, temperature ranging from 30 to 35°C is suitable for green leafs. Similarly, for tomato, a different range of temperature and humidity is required.

Advantages	Disadvantage
▪ Soilless ▪ Sunlight not required ▪ Easy maintenance ▪ No climate effect ▪ Optimal water conservation ▪ New age of natural farming ▪ Facilitate organic farming	▪ Limited spaces ▪ Required skilled labor
Future scope ▪ Power consumption could be reduced in future ▪ Scheduling of low power ▪ Integration with advanced techniques like LoRaWAN ▪ Renewable energy sources Implementation using solar and wind	Challenges ▪ Limited number of crops can be cultivated using vertical farming ▪ Slow-growing plants like rice, paddy cannot be cultivated ▪ Deep-rooted plants like potato are not suitable ▪ High space–requiring plants such as corn is avoided in vertical farming ▪ Lighting is another issue in this for photosynthesis purpose; plants require light with suitable amount of intensity (lux: unit of light intensity measurement)

To control these parameters, we have used a fully autonomous IoT-enabled architecture equipped with sensor module and actuators. Further to enhance crop yield, we have implemented machine learning algorithm to maintain all parameters at certain threshold for suitable growth of the plant.

1.7 ADFC ANALYSIS

An overall analysis is show cased in the following table.

1.8 CONCLUSION

VF is the new age of farming. The chapter provides an exclusive idea to establish VF in integration with cutting-edge technologies like IoT and machine learning. An indoor VF ecosystem has been elaborated in detail. Further deploying sensor modules and collecting data from ecosystem have been simulated on Cisco Packet Tracer to access the information remotely. The architecture for LoRa and IoT networks has been elaborated in this chapter suitable for VF. We have discussed various approaches for integrating LoRa into IoT to develop innovative solutions for smart farming, etc. The challenges and opportunities have been explored in ADFC chart in the chapter.

1.9 FUTURE SCOPE

With revolution of technology, farming is also witnessing a new dimension. In coming future natural farming is going to next era with technology. In near future precision agriculture is going to be adapted in large scale with the help of IoT.

KEYWORDS

- **vertical farming**
- **IoT**
- **AI**
- **machine learning**
- **sensor nodes**
- **hydroponic**
- **aeroponic**
- **sustainable farming**

REFERENCES

1. Jürkenbeck, K.; Heumann, A.; Spiller, A. Sustainability Matters: Consumer Acceptance of Different Vertical Farming Systems. *Sustainability* **2019,** *11* (15), 4052.
2. Butturini, M.; Marcelis, L. F. Vertical Farming in Europe: Present Status and Outlook. In *Plant Factory*; Academic Press, 2020; pp 77–91.
3. Beacham, A. M.; Vickers, L. H.; Monaghan, J. M. Vertical Farming: A Summary of Approaches to Growing Skywards. *J. Horticul. Sci. Biotechnol.* **2019,** *94* (3), 277–283.

4. Zambon, I.; Cecchini, M.; Egidi, G.; Saporito, M. G.; Colantoni, A. Revolution 4.0: Industry vs. Agriculture in a Future Development for SMEs. *Processes* **2019**, *7* (1), 36.

5. Kalantari, F.; Tahir, O. M.; Joni, R. A.; Fatemi, E. Opportunities and Challenges in Sustainability of Vertical Farming: A Review. *J. Landscape Ecol.* **2018**, *11* (1), 35–60.

6. Benke, K.; Tomkins, B. Future Food-Production Systems: Vertical Farming and Controlled-Environment Agriculture. *Sustain.: Sci., Pract. Policy* **2017**, *13* (1), 13–26.

7. Kozai, T.; Niu, G.; Takagaki, M., Eds. *Plant Factory: An Indoor Vertical Farming System for Efficient Quality Food Production*; Academic Press, 2019.

8. Ho, T. T.; Shimada, K. The Effects of Climate Smart Agriculture and Climate Change Adaptation on the Technical Efficiency of Rice Farming—An Empirical Study in the Mekong Delta of Vietnam. *Agriculture* **2019**, *9* (5), 99.

9. Khoa, T. A.; Man, M. M.; Nguyen, T. Y.; Nguyen, V.; Nam, N. H. Smart Agriculture Using IoT Multi-Sensors: A Novel Watering Management System. *J. Sensor Actuator Netw.* **2019**, *8* (3), 45.

10. Jagustović, R.; Zougmoré, R. B.; Kessler, A.; Ritsema, C. J.; Keesstra, S.; Reynolds, M. Contribution of Systems Thinking and Complex Adaptive System Attributes to Sustainable Food Production: Example from a Climate-Smart Village. *Agric. Syst.* **2019**, *171*, 65–75.

11. Trilles, S.; Torres-Sospedra, J.; Belmonte, Ó.; Zarazaga-Soria, F. J.; González-Pérez, A.; Huerta, J. Development of an Open Sensorized Platform in a Smart Agriculture Context: A Vineyard Support System for Monitoring Mildew Disease. *Sustain. Comput.: Inf. Syst.* **2019**.

12. Al-Chalabi, M. Vertical Farming: Skyscraper Sustainability? *Sustain. Cities Soc.* **2015**, *18*, 74–77.

13. Beacham, A. M.; Vickers, L. H.; Monaghan, J. M. Vertical Farming: a Summary of Approaches to Growing Skywards. *J. Hortic. Sci. Biotechnol.* **2019**, *94*(3), 277–283.

14. Roberts, J. M.; Bruce, T. J.; Monaghan, J. M.; Pope, T. W.; Leather, S. R.; Beacham, A. M. Vertical Farming Systems Bring New Considerations for Pest and Disease Management. *Ann. Appl. Biol.* **2020**.

15. Jürkenbeck, K.; Heumann, A.; Spiller, A. Sustainability Matters: Consumer Acceptance of Different Vertical Farming Systems. *Sustainability* **2019**, *11*(15), 4052.

16. Tzounis, A.; Katsoulas, N.; Bartzanas, T.; Kittas, C. Internet of Things in Agriculture, Recent Advances and Future Challenges. *Biosyst. Eng.* **2017**, *164*, 31–48.

17. Botta, A.; De Donato, W.; Persico, V.; Pescapé, A. Integration of Cloud Computing and Internet of Things: A Survey. *Future Gen. Comput. Syst.* **2016**, *56*, 684–700.

18. Mekki, K.; Bajic, E.; Chaxel, F.; Meyer, F. A Comparative Study of LPWAN Technologies for Large-scale IoT Deployment. *ICT Exp.* **2019**, *5* (1), 1–7.

19. https://www.microchip.com/wwwproducts/en/ATmega328

CHAPTER 2

Applications of IoT in Agriculture

P. S. RANJIT[1*], G.S. MAHESH[2], and M. SREENIVASA REDDY[1]

[1]*Aditya Engineering College (A), Surampalem, Andhra Pradesh, India*

[2] *Sri Venkateswara Engineering College, Tirupathi, Andhra Pradesh, India.*

Corresponding author. E-mail: psranjit1234@gmail.com

ABSTRACT

The agricultural industry must rise to meet demand, regardless of environmental issues such as unfavourable weather and climate change, in view of an increasing population, currently expected to hit 9.6 billion by 2050. Agricultural industries would need to use emerging technology to gain an advantage in order to meet the needs of the increasing population. Increased operating productivity, lower costs, reduced waste and improved quality of the yield of new agriculture applications for IoT and smart agriculture would help. In these lines, make use of wireless sensor networks in agriculture as well as its difficulties in implementation in agri-environment and use of IoT and its use in agri-sector as well as biofuels through smart agriculture were revealed.

2.1 INTRODUCTION

The worldwide populace could hit 8 billion by 2025 and more than 9.5 billion in next 50 years.[1,2] The constantly developing total populace's wellbeing and prosperity depend fundamentally on adequate quality and supply nourishment. Other than expanding worldwide interest for nourishment, individuals need sound nourishment things at low costs, however much as could reasonably be expected. These realities make the Agri-nourishment industry probably the biggest business on the planet. The significant changes in the farming and nourishment industry were propelled by populace development, urban advancement, atmosphere, and monetary improvements. Development and progress of the technology is an incredible mode for the best possible distribution of land necessary for homes and agriculture. With the proficient utilization of the constrained assets of the earth, the generation of expanded amounts of sound nourishment at sensible costs places incredible weight on the worldwide Agri-nourishment advertisement. Nearby expanded nourishment needs and deficiency of arable land and water supplies, just as environmental change, are the greatest difficulties confronting reasonable farming and the nourishment business in the 21st century[3] All Agri-nourishment organizations, governments, and organizations should try sincerely and team up so as to accomplish feasible agribusiness and nourishment business and to satisfy the need that creation of nourishment should double by 2050. The top needs for present-day society are changes to human wellbeing and prosperity and improved nourishment security and natural manageability. Expanded amount of greater nourishment delivered in a practical way relies upon various factors, for example, social, financial, specialized, and political elements. In the present nourishment industry, the continuous and fast changes in customer requests, the abbreviated lifecycle of nourishment items, pressure on time to showcase, guidelines on wellbeing, and so forth are largely difficulties. Nourishment security alludes to nourishment verified by nourishment creation, fabricating, taking care of, dissemination, stockpiling, and utilization from all types of poisons. Every single person ought to consistently be given proper measures of cleaner, better, and more secure and nutritious nourishment consistently.

New nourishment products of upgraded esteem are the aftereffect of innovative advances and their incorporation in farming and the nourishment part. All in all, innovation has made nourishment fabricating, stockpiling, appropriation, and warehousing snappier and increasingly successful and progressively proficient. Therefore, the number, assortment, and nature of items in nourishment items that meet the wellbeing and healthful prerequisites of shoppers seem to have expanded. Nevertheless, information and communications technology were only accessible in the horticulture and food field due to the aim of Internet of things (IoT).

2.2   WIRELESS SENSOR NETWORK

Detailed schematic diagram to make use of wireless sensor network in day to day life is revealed in the Figure 2.1. Innovative propels have permitted the making of tangible gear, from basic detecting hardware to quantify an intrigue parameter, to create some sort of sign yield, and to generation of a lot of little, low-cost, low-control, and smart sophisticated gadgets. Such smart systems, known as sensor nodes, can follow a wide exhibit of parameters in an assortment of utilizations and receive and transmit the information more extensively utilizing remote innovation (e.g., Bluetooth, Zigbee), remote neighborhood (WLANs), or long-separation (e.g., GSM/GSM/GSM/GSP). The variety of sensor hubs shaping a remote correspondence organize alludes to the wireless sensor network (WSN). WSN leads constant observing and assortment of information of intrigue, in this manner permitting the procurement of information required for specific acts. The WSN applications are differing and essentially all around the globe: producing, military gadgets, wellbeing and security, transport and coordination, astute homes, amusement, social insurance at home, natural sciences, agribusiness, and so forth. For dangerous and remote natural checking, the advantages of WSN are particularly watched.

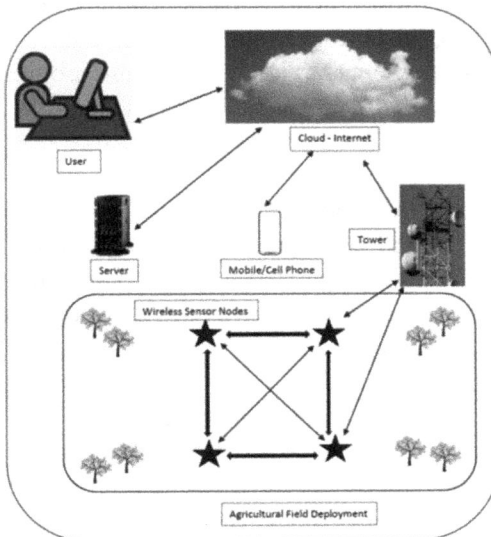

FIGURE 2.1   Wireless sensor network.[4]

2.3   WSN IN THE AGRI-NOURISHMENT DIVISION

The Agri-nourishment division can be viewed as altogether not quite the same as other field fields of WSN. The observation of the atmosphere and plants, just as nourishment detectability, is of prime enthusiasm for request to boost Agri-nourishment results. Detailed perception about make use of wireless network in the agri-environment was shown in Figure 2.2. Some issues like, plant culture, soil type, variation in atmosphere along with detecting the nourishment from the available crops and providing territories based on these issues is slightly a challenging task for WSN in agribusiness. By the bye, WSNs are regularly utilized in the Agri-nourishment field.[5,6] Information transmission to the remote control focuses on the following tasks, including different remote detecting hubs and consistent estimation of different parameters continuously (e.g., air and soil temperatures, stickiness, wind speed, course, precipitation, daylight, creature wellbeing status, and nourishment generation status). In light of the information gathered and examined, remote clients can settle on fitting and convenient choices and activities that help in the enhancement of Agri-nourishment generation. Innovative progresses permit the checking of creature recognizability and the discernibility of nourishment remotely and in Echtzeit utilizing an assortment of detecting components by means of WSN and remote advancements in natural and horticultural conditions. The constant and top-to-bottom assortment and assessment of gathered information empower ranchers to have significant, progressively, skill and sufficient treatment at the ideal spot. This has added to a fantasy of horticulture with accuracy. An assortment of cutting edge sensors is utilized to deliver exactness horticulture.

Utilizing the Global Positioning System (GPS) and Global Navigation Satellite System (GNSS) permits continuous observation of territorial climatic conditions, the dirt's properties, (e.g., compaction, ripeness, heat), plant status, (e.g., water quality, potential maladies), creature wellbeing, and so on. Fitting utilization of soil, antimicrobials, herbicides, composts, and pesticides adds to maintainable Agri-nourishment generation. The observation of nourishment quality and wellbeing in all phases among ranchers and shoppers was permitted by methods for a peruser's radio frequency ID (RFID) gadget to receive information from a transponder (or a tag) onto a radio recurrence wave, and the transmission of information through reception apparatuses to a PC associated with a nearby system or the web to complete a compelling methodology. Accordingly, RFID and WSN all in all characterize nourishment items with different other information.[7,8] RFID labels are utilized for observing and following animals, alongside their utilization in nourishment discernibility frameworks.

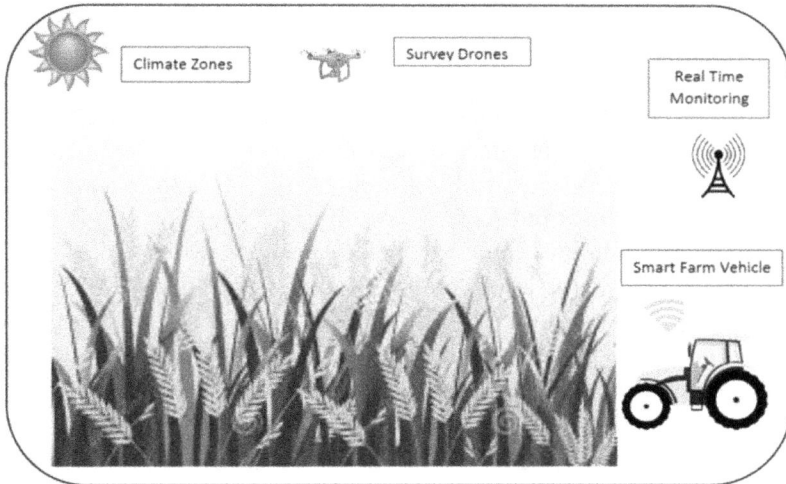

FIGURE 2.2   WSN in agri environment.[9]

2.4   DIFFICULTIES FOR WSN APPLICATIONS IN THE AGRI-NOURISHMENT SEGMENT

In view of the many advantageous conditions of WSN in the agricultural and food processing sectors, many obstacles still exist for WSN to demonstrate its maximum capacities in the agri-food field. The dynamic system topology is required for WSN applications in farming and nourishment generation. WSN should likewise be secure, simple to utilize, and simple to keep up under unstable or unforgiving natural conditions.[5] Low WSN costs are one of the key drivers for WSN's wide utilization of Agri-nourishment. Furthermore, WSN utilized in unpredictable or outrageous natural conditions must be secure, simple to utilize, and simple to keep up. One of the key factors in the huge usage of WSN in farming and nourishment industry is the minimal effort of WSN. Innovative progresses and diminished mechanical expenses enable this to occur. A power module (battery) is the weakest purpose of a remote sensor hub.

The objective is to extend the life expectancy of the hub and, eventually, an entire WSN in light of the fact that the remote sensor hub has constrained control. Enhancing power utilization through equipment and programming component plan and interchanges conventions is one arrangement. Another approach to keep up a steady vitality source is to receive power from the nature (e.g., from sun-based, wind, and vibratory vitality assortment frameworks). When WSN assembles a colossal measure of heterogeneous data, the information executives perform a significant test. For information transmission from WSN to a remote control focus, more prominent data transfer capacity is required. This can be overwhelmed by permitting prepreparing data to expand control utilization at the sensor hubs.

The nearby administration of heterogeneous information gathered continuously is a major issue for low-vitality remote sensor hubs at the end of the day. In perspective on the way that arranging, breaking down, capacity and reproduction of an enormous volume of an assortment of information is critical to WSN's effective execution in the Agri-nourishment advertisement; different information mining innovations, computerized reasoning, and measurable examination apparatuses are utilized to recover helpful data from countless various information gathered by a few wires in the objective region. Joining helpful data, chronicled information and rural ability gathered, reinforcing frameworks of the board and basic leadership, adding to economical farming and nourishment generation. Finding a harmony between the control utilization and information executives is as yet a significant test in the exploration region of the WSN. In any case, WSN has just indicated the immense potential for upsetting everything from nourishment creation, producing, shipping, conveyance, stockpiling, and intending to selling and exchange agribusinesses and nourishment industry. The goals of the issues that the WSN is looking in the nourishment business are unquestionably adding to reasonable horticulture and the nourishment business, joining it with new innovations, for example, IoT.

2.5   INTERNET OF THINGS (IOT)

A very new worldview, that is IoT, has risen up out of WSN usage and RFID availability of keen gadgets, machine-to-machine association, link systems, and internet applications.[11] The IoT is a system of items and people related as shown in Figure 2.3 as it can be accessed and controlled and monitored from any place (Cortés 2015). Shrewd items can be characterized (by a virtual name), cooperated (remotely and as specially appointed interconnected articles), and draw in (by detecting) with nearby situations as key building squares of IoT. Since its presentation, IoT has become an extensive and inescapable reinforcing our dailies and driving our economy, in various applications and as a quickly developing advancement. All positive IoT executions[13] ought to be fruitful paying little mind to the usage situation: comprising ubiquitous systems, utilizing conventions in a similar language, and conveying without the interest of people. Such savvy gadgets remotely, safely, and dependably receive objective data. These must likewise be associated with the web, which can be recovered and handled progressively for a mix of the gathered information.

FIGURE 2.3   Internet of things.

Apply design acknowledgment calculations and apparatus learning procedures for the examination of the information accumulated and for the extraction of proper and separating data. Trades information in a sheltered and secure manner by means of remote system imagines gigantic information volumes and results in a simple-to-utilize and adaptable interface for the individuals who can create constant info and in this way moves proactive conduct to solid, prescient, and preventive advances.

2.5.1 IOT IN THE AGRI-NOURISHMENT DIVISION

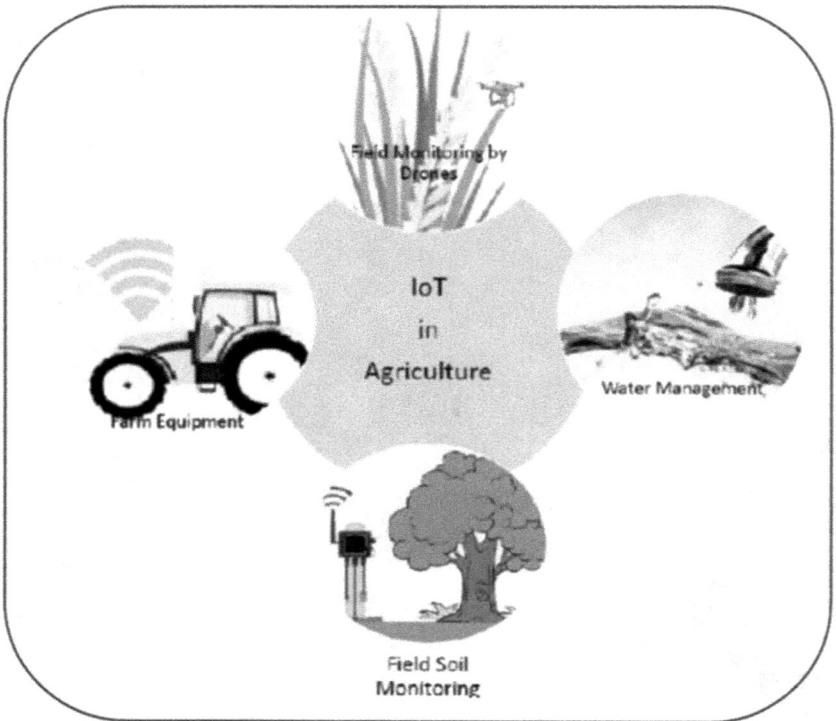

FIGURE 2.4   IoT applications in agri-nourishment.[9]

Among the numerous zones where IoT has indicated its guarantee, farming can be viewed as one of the quickest developing and most neglected by IoT advancement as shown in Figure 2.4. Because of the mechanical improvements, the development, get together, preparing, transportation, appropriation, and capacity are essentially upgraded, both essential and optional. Totally shut source and open-source programming to business IoT applications permit end clients to alter them for their needs and inclinations. Sending, designing, running, overseeing, and keeping the framework straightforward, financial savvy must be authorized. In mechanical improvement, for instance, conservative, scaled-down, proficient, easy-to-understand, and reasonable frameworks are created and utilized for quick on location nourishment quality checking by recognizing an objective material. These frameworks are associated with advanced cells and are a piece of an IoT dream. The general biosensor and spectroscopy-centered demonstrative frameworks[14] can assemble enormous volumes of information from which valuable data is acquired utilizing Fog and Cloud number crunchers. The majority of the difficulties confronting these applications incorporate example arranging, exactness of estimations, and utilization of conventions. Various sensors have been utilized, going from temperature, moistness, water contamination, air contamination, pressure, optical sensors, computerized (e-language/e-nose) compound items that distinguishes synthetic substances based on smell/smell, to remote sensors in the Agri-nourishment industry. Utilized as a major aspect of an IoT biological system in the nourishment business, these gadgets accommodate observation of different parameters over the nourishment generation chain, including natural checking, soil/plant status and creature wellbeing, remote location, and following of nourishment items.

For example, sensor data that determine the proximity of heat, potassium, sodium, and others to the soil will activate the smart watering arrangement and enable for the optimal use of agrochemical goods. The IoT vision can likewise be utilized to furrow fields and reap crops with mechanical tractors. Information from various sensors can be consequently transmitted to the individual tractor to figure out where and when to plant. Smart sensors and independent frameworks were created to make a differentiation among seed and weed and to viably annihilate weeds. The utilization of cameras and hyper-phantom frameworks can give a progressively definite knowledge into the strength of yields and conceivable harm. Based on these outcomes, automatons can likewise be utilized in the proper plant at the correct time of development to supply composts, weed executioners, and different agrochemicals. Recognizing or disposing of the debasement of nourishment items is necessary so as to keep away from any further undesired results.

The IoT-wearables consequently keep away from cross-pollution among representatives and merchandise for the motivations behind constant observation of the wellbeing and cleanliness of nourishment-handling laborers. In the nourishment retail process, a similar methodology is fundamental. Likewise, IoT gives continuous, progressing knowledge into the states of air during transport and conveyance and can regulate cargo development through cameras as well as satellites. On account of the IoT model, the condition of the nourishment item during shipment and preparing, is slightly open ended task needs to be addressed. Obviously in the Agri-nourishment industry, the acquaintance of IoT contributes with a tremendous measure of heterogeneous information gathered continuously all through the entire nourishment inventory network. The control of these data is viable in tending to the limitations of transfer speed, idleness, availability help, and acknowledgment position.

It implies that distributed computing is not customized for regions, for example, ranger service and the nourishment business, which are huge scale and comprehensively scattered. Actualizing the Fog organize, a delegate layer between end clients and the cloud, permits an increasingly effective, exact, and speedy reaction framework for neighborhood stock-piling and preparing. What IoT innovation in agribusiness is chosen and sent depends, be that as it may, on what buyers need to follow and what moves they make. A definitive presentation of the framework circumstance will be affected by choosing the privilege IoT stage for a specific application. Customary farming has advanced to an amazingly serious and exact development, with a high profitability and proficiency together with decreased inputs, utilizing the guide of IoT innovations. This objective is cultivated by the structure of an interoperable IoT-eco-arrangement of a few heterogeneous systems, offices, and usage (gadgets, identifiers, gear, and so forth). All features of interoperability additionally must be taken up: utilitarian (framework to-framework information trade), syntactic (information group), etymological (human-accommodating portrayal and cognizance of substance), and managerial (information for framework-to-topographical data exchange). The data on the name shows the type of staple, the ranch of source, and different subtleties of significance in the rest of the nourishment store network to customers and different partners. In the event that the faulty material prompts solid, quality nourishment on the client plate, self-governing frameworks incorporate computerized thinking. So as to gather, aggregate, disperse, utilize, and share dependable and significant data, policymakers settle on right speculation choices in the rural and nourishment industry which are the best approach to make the change to economical horticulture and the nourishment advertise. It might likewise be the plot that IoT licenses access to enormous-scale information identifying with the way of time and cost sparing through on-schedule and exact choices and activities where, how, through whom nourishment is prepared, conveyed, disseminated, and expended. The trading of data, coordinated effort, and collaboration between the different on-screen characters in the general nourishment process in this way reinforce the avocation for horticulture and nourishment creation. Consequently, new items and procedures have been built up that fulfill client needs and requests and alter fabricating levels to advertise patterns.[15] So as to accomplish higher caliber and increasingly focused farming and nourishment generation, numerous IoT-related issues, for example, reconciliation forms, institutionalization, wellbeing, financial issues, advertise appropriation, and so on must be effectively tackled. All the more explicitly, in all nourishment generation, stockpiling, and transport stages, security and protection issues should be taken care of in a way that lone approved shoppers are permitted to get to delicate subtleties and to perform proper acts.[16]

2.5.2   *DIFFICULTIES FOR IOT APPLICATIONS IN THE AGRI-NOURISHMENT INDUSTRY*

New IoT-based answers for Agri-nourishment are continually making progress to meet the prerequisites of the present rehearses, because of quick innovative advancement. The Agri-nourishment industry is certainly upset by powerful IoT innovations. By and by, IoT sending in farming and the nourishment business faces different obstacles and difficulties. Web access is compulsory for IoT execution at any scene and for nothing out of pocket. A definitive IoT inescapability relies upon the creation of modest astute detecting frameworks, adaptability, lower vitality utilization, expanded handling limits, issue affectability, and a situation of trust. The advancement and development in execution of IoT innovation require consistent and complex framework, arranged elements distinguishing proof, security and protection, norms and guidelines, and administration.

Because IoT relies on a number of heterogeneous steps, IoT is concerned with problems and basics, information from managers and the use of vitality. Sensor modules are utilized in horticulture in the provincial, to a great extent uncheckable atmosphere, so they must be watched from natural impacts (e.g., temperature vacillations, solid breezes, heat, high mugginess, vibration, and low sun-powered radiation) and must be steady and long-haul tough. Sensor modules sent at the objective site as gadgets with a restricted power ought to be planned and customized to meet lifetime prerequisites. A few rest calculations and vitality stockpiling frameworks (e.g., sun-powered boards and wind turbines). This situation might be tackled. The unwavering quality of cooperation between sensor hubs likewise relies upon unforgiving ecological elements. To conquer this issue, proficient and dependable innovation should be utilized, and information stockpiling, information preparing, information recuperation openings, and so forth should be focused on. What's more, joining and execution of a portion of the remote advancements frequently utilized depend on a few sensor hubs, their position at the objective site, the area of the receiving wire, and the recurrence of activity. Vitality productivity, organizing highlights, adaptability, and heartiness should likewise be focused. In the Agri-nourishment industry, extra IoT utilization issues incorporate the discovery and the board of an expansive scope of sensor hubs. Absolute IPv6 Addressing Scheme execution will assume a significant job in taking care of this issue. By and by, it is a genuine and moving undertaking to exist together and interoperate in one system of gadgets with different capacities (e.g., control supply, PC control, and fringe gadgets). Heightened research and work in IoT are supported by the association and interconnection of different sensor modules, climate stations, RFID frameworks, vehicles and hardware, web passages, cell phones, tablets, and different gadgets. Norming basic IoT advances and conventions for availability (e.g., Zigbee, Wireless HART, BLE, Lora, DASH7, and Wi-Fi) permits a total IoT system of an assortment of gadgets. Information the board is another major issue in IoT vision. With the goal for clients to gather helpful data, suitable information must be gathered, and an enormous and heterogeneous information assembled continuously should be separated, handled, put away, dissected, and imagined. Distributed computing as well as Fog figuring in IoT give basic preparing and capacity assets for information treatment of the IoT gadget. The principle issues with IoT and cloud mix anyway concern institutionalization, adaptability, utilization of intensity and assets, security, and protection. Security affirmation and obscurity and approved access at different IoT levels are mandatory in the rural and nourishment division to embrace IoT arrangements. The distinguishing proof of mystery examples and assortment of valuable data, information creation and sharing can likewise be practiced utilizing a scope of information mining strategies, fake intelligence, neural systems, huge information and investigation, and other basic leadership procedures. The completely mechanized procedure is applied along these lines. The IoT at that point coordinates and advances distinctive business forms by straightforward data stream between all partners in the IoT model. Notwithstanding these obstructions, the powerlessness of individuals to know and to acknowledge IoT arrangements dependent on cultivating, business people, and restricted access to ICT foundation in different districts ought not to be ignored.

2.6 INFLUENCE OF SMART AGRICULTURE (SA) ON BIOFUELS

Since 2000, more than 500 per cent has increased in the creation of global biofuel to replace the conventional fossil fuels. By 2020, creation is required to additionally increment by half, and by 2050. Strategies have assumed a basic job in quickly expanding the creation of liquid biofuels during the last 5 or 6 years, fundamentally as far as transport. The longing to improve security in vitality, decrease outflows of green house gases (GHGs), cultivate provincial advancement, and increment ranch salaries roused the strategy support for biofuels.

With the quickened presentation of new and broadened motivators, a superior proof base is currently accessible that can break down the effects of expanded generation of biofuels and decide if approaches can be changed to meet developing needs and concerns.

2.6.1 A FEW POTENTIAL BIOFUEL SUPPORTERS OF THE SA TARGETS ARE RECORDED UNDERNEATH

Bio-energizes can add to improving access to current residential and beneficial vitality benefits that add to economical increments in efficiency and pay. An ongoing investigation of little scale activities in bioenergy (FAO, 2009) has indicated that this upgrade can be accomplished with least supportability perils.

Biofuels can improve environmental change flexibility and change-ability, especially in little scale creation. In any case, by making a connection between vitality security and yield returns, they can likewise make their own atmosphere dangers. This hazard is particularly high if the decent variety of feedstock is low. The effect of carbon sequestration and GHG discharges is increasingly unpredictable and a subject of a lot of exchange.

As biomass production by photosynthesis consumes a similar amount of CO_2, emitted through biofuel use, bioenergy is widely seen as CO_2-unbiased. The connections of, for example, nitrogen, phosphorus, and water cannot be muddled over between the carbon cycle and other characteristic cycles. These components are also essential for photosynthesis and are consumed by delivering biomass. Soil additives are evacuated and should later be included. This adds can add to GHG contamination, especially nitrous oxide (e.g., application to composts). An intensive appraisal of life cycles that can consider crop generation and dispersion and consider changes in immediate and aberrant land utilization should be done.

Some prescribed procedures that can support the yields of biofuels for moderation of environmental change incorporate: agroecological zoning so as to counteract high carbon zones from creating biofuels (e.g., essential timberlands and peatland). The use of build-ups to manufacture biofuels has kept them from impacting their use as soil by management or as feed for livestock. And agribusiness is a low-carbon cultivation activity that can sequestrate carbon even now and then.

All the more explicitly, biofuel approaches and projects, instead of measures that intentionally advance the requirement for biofuel, should work pair with rural improvement programs. To decrease dangers and adventure the open doors for bioenergy improvement, a sound and coordinated way to deal with biomass vitality, specifically the advancement of biofuels, is required. An exhaustive comprehension of the circumstance, the related advantages and difficulties, just as the chance and the exchange off encourages a political and administrative system that depends on sound and versatile enactment (for instance, objectives and motivations) and methods for authorizing these strategies;

2.6.2 PUSHING AHEAD—CONCEIVABLE VITALITY ANSWERS FOR SA

FAO projections for 2030 show that the extent of hand-developed and creature fuelled land in creating nations will decay. This move in cultivating rehearses offers horticultural efficiency and drudgery open doors for ranchers. In any case, poor ranchers regularly cannot bear the cost of costly hardware and gear. It is important to give creative business and network demonstrating that enables smallholder ranchers to get to better innovation, (e.g., rental frameworks or cooperatives). The change to exceptionally motorized rural frameworks is probably going to diminish farming work and openings for work in country zones. Well-structured techniques and projects, in the horticultural worth chain and in other non-rural country occupations, must give options in contrast to employments.

2.6.3 ADVANCES FOR VITALITY SAVVY NOURISHMENT AND SA

The slow move to vitality shrewd nourishment frameworks needs a mix of appropriate vitality innovations, gear, and offices in cultivating networks. There are a scope of improvements which can be a piece of intensity educated nourishment frameworks, for example, wind turbines, sun-based boards, photovoltaic boards, biogas-handling machines, power genera-tors, bio-oil extraction and refinement offices, ethanol creation aging and refining frameworks, pyrolysis units, water siphons with aqueous vitality change instruments, wind and bio-vitality water siphons. The innovation adds to the creation of crude materials close to the source. It is hard to distinguish vitality savvy sustenance's problem areas' and activity objec-tives with information presently available. Different evolved ways of life experience very mind boggling forms that require various types of vitality inputs. In summary, more investigations will be needed into the ties between use of fertility, yields, and cost of production in different rural areas. In working and development field efficiencies can be up to 90%, in treatment and in grain harvests 65–70%. In any case, yields and plot sizes rely upon execution. The generally fuel utilization is 600–1200 MJ/h, for moldboard furrowing (MJ/ha), 200–4900 MJ/ha disking and 80–160 MJ/ha for manor and 150–300 MJ/ha for the use of smelling salts.

Agribusiness frameworks with ordinarily low vitality necessities and broad territories of horticulture and touching, for example, those for Australia or NZ, can work with a vitality utilization of up to a few gigajoules per hectare. In nations, for example, the Netherlands or Israel, the vitality interest for input-serious agribusiness can reach up to GJ/ha of 70–80. In calories, China is currently more vibrant than the United States or the European Union with a huge plant percentage, large water system and severe preparation, as proposed by the horticultural market. Nitrogen (half got from inorganic composts) provided about 60% of the supplement in crop development following the 1978 horticulture changes in China. Over 80% of the national interest for protein was separated from crop generation.

The agribusiness business depends vigorously on non-renewable energy sources however in crowded areas it has had the option to encourage about 8.5 individuals per ha and up to 15 individuals. This finding is likewise credited to a creature protein provincial eating regimen. Wasteful utilization of nitrogen prompts regular misfortunes in abundance of half and can some of the time be 60–70% all things considered. From various perspectives, upgrading manure yield to accomplish most extreme plant development with low information sources will altogether improve the nourishment generation vitality balance. It would likewise add to secure the earth and decrease ranchers' costs. Regardless, it might be fitting to utilize more manure to build yield in specific territories, as in Africa, to arrive at most extreme vitality proficiency in nourishment generation. A further huge technique for diminishing manure exhaustion could be checking soil disintegration.

For horticulture, water quality has become a problem. By and by, it might require more vitality to accomplish more prominent water system yield. For instance, trickle water system, which expands water effectiveness, expects vitality to push the water. A ton of the waste control is regularly utilized for drying. In remote zones, expanding water system requires vitality innovation, for example, sun-powered controlled siphons to spare manual physical work in rustic regions of the vitality matrix. Water system execution can be up to 95%; the normal paces of yield for good field tasks is between 65 and 75%, while the effectiveness of the wrinkling can be just 3–40%. Asia can hypothetically significantly increase water system limit. For soil readiness, fluid powers are typically required. The measure of vitality required for this is subject to the climate, soil compaction, and different variables (wet or dry soils). In a yield stage, the most vitality expending process is land lying, especially furrowing. As an outcome, diminished culturing crop frameworks, particularly no-till frameworks in the midst of high vitality costs, turned out to be especially appealing. There are different innovative answers for diminish the utilization of intensity. Such strategies limit the moving power and the slippage of blended reapers (e.g., the advancement of tractor pneumatic frameworks). A significant zone of intercession is additionally vitality preservation in nurseries, domesticated animals, and ranch structures. The better utilization of warmth siphons (for the most part of the water driven type of pressure controlled by electric engines) and warmth recuperation frameworks could decrease vitality use. The dehumidification offices and cooling can likewise be given by both. Air-to-water heat siphons and water-to-water heat siphons can fundamentally expand vitality productivity in all tasks including power, likely combined with geothermal power sources. Pipeline ventilation, warmed floors, electric inflator, and water are altogether specialized arrangements which can be thought about. Probably the most prudent vitality productive measures are legitimate development, protection and ventilation of structures and nurseries.

For all activities that structure some portion of the agri-food chain, the great and most noticeably terrific gauge of the vitality force per shopper unit can be made. Such rehearses are not part of the IPCC's GHG program however are illustrative of the Industrial Processes and vitality industry. Or maybe, they are managed by the GHG.

KEYWORDS

- **wireless sensor network**
- **internet of things**
- **smart agriculture**
- **agri-nourishment**
- **biofuels**

REFERENCES

1. Tzounis, A.; Katsoulas, N.; Bartzanas, T.; Kittas, C. Internet of Things in Agriculture, Recent Advances and Future Challenges. *Biosyst. Eng.* **2017,***164*, 31–48.
2. Singh, R. L.; Mondal, S. *Biotechnology for Sustainable Agriculture—Emerging Approaches and Strategies*; Woodhead Publishing, Elsevier, 2018.
3. Chen, H.; Yada, R. Nanotechnologies in Agriculture: New Tools for Sustainable Development. *Trends Food Sci. Technol.* **2011**, *22*, 585–594.

4. Ojha, T.; Misra, S.; Raghuwanshi, N. S. Wireless Sensor Networks for Agriculture: The State-of-the-Art in Practice and Future Challenges. *Comput. Electr. Agric.* **2015,** *118,* 66–84.

5. Wang, N.; Li, Z. Wireless Sensor Networks (WSNs) in the Agricultural and Food Industries. *Robot. Auto. Food Ind.—Curr. Future Technol.* **2013,** 171–199.

6. Srivastava, A. K.; Dev, A.; Karmakar, S. *Nanosensors and Nanobiosensors in Food and Agriculture.* Environmental Chemistry Letters Springer International Publishing AG, 2017.

7. Alfian, G.; Rhee, J.; Ahn, H.; Lee, J.; Farooq, U.; Fazal Ijaz, M.; et al. Integration of RFID, Wireless Sensor Networks, and Data Mining in an e-Pedigree Food Traceability System. *J. Food Eng.* **2017.**

8. Sethi, P.; Sarangi, S. R. Internet of Things: Architectures, Protocols, and Applications. *J. Electr. Comput. Eng.* **2017,** 25.

9. Team DF. IoT Applications in Agriculture—4 Best Benefits of IoT in Agriculture, 2018. https://data-flairtraining/blogs/iot-applications-in-agriculture/.

10. Internet of Things. https://pixabay.com/illustrations/iot-internet-of-things-internet-4085382/.

11. Atzori, L.; Iera, A.; Morabito, G. Understanding the Internet of Things: Definition, Potentials, and Societal Role of a Fast Evolving Paradigm. *Ad Hoc Netw.* **2016.**

12. Cortés, B.; Boza, A.; Pérez, D.; & Cuenca, L. Internet of Things Applications on Supply Chain Management. *Int. Sch. Sci. Res. Innov.* **2015,** *9,* 2493–2498.

13. Maksimovi´c, M.; Vujovi´c, V. Internet of Things Based e-health Systems: Ideas, Expectations and Concerns. In *Handbook of Large-Scale Distributed Computing in Smart Healthcare, Scalable Computing and Communications;* Springer International Publishing AG, 2017; pp 241–279.

14. Rateni, G.; Dario, P.; Cavallo, F. Smartphone-based Food Diagnostic Technologies: A Review. *Sensors* **2017,** *17,* 1453.

15. King, T.; Cole, M.; Farber, J. M.; Eisenbrand, G.; Zabaras, D.; Fox, E. M.; et al. Food Safety for Food Security: Relationship between Global Megatrends and Developments in Food Safety. *Trends Food Sci. Technol.* **2017.**

16. Tewari, A.; Gupta, B. B. Security, Privacy and Trust of Different Layers in Internet of Things (IoTs) Framework. *Future Gen. Comput. Syst.* **2018.**

17. IoT World Today, 2018. https://www.iotworldtoday.com/webinar/why-open-source-works-for-renewable-energy/: influx data.

CHAPTER 3

IoT-Based Smart Irrigation Systems for Smart Agriculture

P.S. RANJIT[1*], B. VIDHEYA RAJU[1], G.S. MAHESH[2], and
M. SREENIVASA REDDY[1]

[1] *Aditya Engineering College(A), Surampalem, Andhra Pradesh, India*

[2] *Sri Venkateswara Engineering College, Tirupathi, Andhra Pradesh, India.*

Corresponding author. E-mail: psranjit1234@gmail.com

ABSTRACT

India's economy depends firmly on agriculture. Around 70% of the inhabitants in India rely upon domesticated animals and a third on fisheries. The water system is a significant determinant in cultivating because it improves rural generation. Ebb and flow water system methods will expand the outstanding task at hand of ranchers, and the harvest will not productively gather water. Overwater and underwater systems are required. Another framework, the Low-Cost Agricultural Field Surveillance System, has like this been built up to unravel these issues.

This strategy brilliantly floods the field by giving water to the plant similarly without the mediation of the maker. This incorporates various highlights, for example, water, heat, downpour detecting, moistness sensor, and temperature sensor. These sensors are utilized at different places on the ranch. The Arduino, which holds ATmega328, kills the engine and utilizes a hand-off to assess the sensor data. For further following the detected qualities and the condition of the engine, it is transmitted to ranchers by means of a Global System for Mobile Communications (GSM) framework and every one of the information is put away in the cloud utilizing the Wi-Fi gadget. The fluid is disseminated equally in the field by this technique. The creation of yields is, in this way, fruitful, and the impacts are beneficial.

3.1   INTRODUCTION

This component involves a dust detector and temperature sensors in the root region of the frame or door to process sensor details and transmit data to the phone using a Global System for Mobile Communications (GSM) unit. A calculation has been worked for the computation of temperature sensor limit data and soil humidity sensors which were modified into a fluid amount control microcontroller (Prateek Jain, 2017). Power board was utilized for photovoltaic. Another reality is that it is conceivable to program information reviews and water system plans on a page, for example, the cell–internet interface.

The programmed framework has been tried for 136 days and sets aside to 90% on a regular water system. More than a year and a half, three digital framework copies are successfully utilized somewhere else. The gadget can be successful in water-confined geologically secluded zones because of its asset freedom and its minimal effort. The soil humidity content was identified in this section utilizing acoustic innovation. The primary motivation behind this methodology is to set up soil humidity estimations progressively. The procedure depends on the connection between sound speed and soil water immersion of two quantities (Prateek Jain, 2017; Wang, 2010). Soil type and humidity content in the soil will vary the sound velocity (G. Yuan 2004).

This section structures a model of a programmed water system framework dependent on sun-oriented vitality and microcontroller. In the zone of paddy, different sensors are mounted. Sensors constantly sense the water level and furnish ranchers with the data by mobile. Without the paddy field, the farmer controls an engine utilizing mobile (Idso, 1981). At the point when the water level hits hazardous levels, the motor is stopped consequently without rancher's compliance.

The Arduino-based mechanization system and the GSM correspondence innovation were utilized. In a specific zone, the water system framework ensures adequate water system continuously. Sensor being art of the frame work, used to check the soil humidity and water level at root level of the plant of the paddy crops. This program is a significant piece of GSM. The framework utilizes GSM to convey. GSM works by means of SMS and is a connection between the processor Arduino and the focal framework. This framework identifies the atmosphere and the states of the field progressively. Such information is transmitted to the customer as SMS and GSM modems utilizing normal AT (Attention) commands (Yuan, 2004; Erdem, 2010). The greater part of GSM capacities is controlled by these directions. The water system timetable of some water system frameworks includes observing the soil, the strain estimation water status with dribble water system by the sandy soil mechanization control framework. For the maker, keeping up the material on location is exceptionally basic. It proposed the structure of a trickle water system gadget concentrated on the Micro-Controller that is an input control framework which tracks and deals with all dribble water system frameworks progressively for more noteworthy adequacy. The board's water system valve allows ranchers to include the right water measurement at an appropriate time, with little regard to accessibility to switch valves (Erdem, 2010; Nemali and van Iersel, 2006). The framework controls valves through an electronic system. So as to set up a productive water system framework on the land with various harvests, some water system systems are utilized (O'Shaughnessy and Evett, 2010).

In any segment of horticulture in India, present-day agribusiness concentrated on nurseries in the ongoing prerequisite. In this invention, the plants' humidity and temperature are exactly regulated. In view of the shifting conditions, sometimes in huge farmhouses, the consistency of all parts of the farmhouse may vary starting with one area then onto the next. The point-by-point data on the water system is utilized for this GSM. The GSM articulation is sent through the android portable application. The joining of programming and equipment gives an improved direction over the conventional manual procedure. The application expects of the utilization of the system both for remote control and for trickle irrigation.

Such sensors give microcontroller values (Davis and Dukes, 2010). Utilizing sequential informing, microcontrollers transmit quality data to the PC. As indicated by sensors progressively, persistent diagram data are demonstrated through the Internet and Android on PCs and Android-based mobiles (Migliaccio, 2010). On the off chance that the sensor data surpass the edge level, the dribble water system parts can be naturally worked through the microcontroller. The Threshold data is put here. The dribble water system can likewise be overseen from anywhere by a portable Android device.

This is an in-house input control framework for following and managing all dribble water system framework tasks in an increasingly compelling way utilizing a small-scale controller controlled trickle water system process (Grant, 2009). Running ON/OFF is a programmed control gadget used to control valves (Chavan and Karande, 2014; Blonquist, 2006). They allow ranchers to use the perfect water calculation at privilege, autonomous from the accessibility of turning valves or engines (Kim, 2008). These reductions stream over wet fields, halting waste at an inappropriate time. It improves crop execution and encourages in all perspectives to spare time. They allow ranchers to use the perfect water calculation at privilege, autonomous from the accessibility of turning valves or engines (Kim, 2008).

The critical wired connectivity system, for example, currently mostly offers knowledge trading between kindergartens and the Fieldbus control framework (Vijaya Kumar, 2011). In these conditions, a totally remote framework can be exceptionally enticing on the off chance that it is to be enhanced with a total remote system (Dalip, 2014; Prakash, 2016). The frame emphasizes that a sensor for temperature and soil humidity can be situated on the appropriate locations to observe soil, the two delicate parameters for plants, with temperature and humidity. (Nallani, 2015). For refreshing the sensor quality used to add data breakdown, a web server has been chosen. It shows that the Internet of things (IoT) is utilized effectively in farming. It shows the utilization of checked and controlled water system frameworks dependent on Arduino and ESP8266, which are both financial effective and simple. Horticulture field reconnaissance system is advantageous to ranchers to water their property helpfully.

3.2 MOTIVATION

The motivation for this part originated from places where the economy depends on cultivating, and the atmosphere conditions add to a lack of precipitation and water shortage. The ranchers in the region depend exclusively on the downpours and the wells to flood the property. Regardless of whether the homestead land has a water siphon, ranchers must run the siphon physically whenever the situation allows. This manual obstruction by the rancher will be decreased by our arrangement. The accompanying points will be served by the Agriculture Field Monitoring System:

- As there is no spontaneous utilization of warmth, a great deal of water has stayed away from being lost.
- Rinsing is the one in particular where there is deficient soil humidity, and the sensors decide if to kill the engine, sparing ranchers a great deal of time.

This additionally permits the ranchers much required rest since they do not need to go and physically switch the siphon on/off.

The developing interest nourishment supplies request a quick improvement in innovation for nourishment creation. In numerous nations where cultivating assumes a significant job for shaping the atmosphere and the earth isotropic, however, we are not ready to utilize agrarian devices to their full degree. The absence of downpour and exhaustion of surface stockpiling assets is one of the fundamental reasons. Customary water gathering from the soil abatements the degree of water, which expanding territories of non-flooded property gradually. Consequently, impromptu water utilization prompts water spillage incidentally. The most significant preferred position of the Agriculture Field Monitoring System is that water is possibly provided if soil humidity is beneath a pre-set edge level.

It is of more vitality for us. Ongoing water system strategies have been utilized by the ranchers by hand search, in which ranchers occasionally inundate the soil, exchanging the water siphon on/off if suitable. Once in a while this procedure devours more water, and now and again the land is postponed by the drying of the harvests. Water lack harms the development of plants preceding unmistakable shrinking. Other than this moderate pace of development, lightweight organic product is a result of water lack. This inquiry can be amended totally by utilizing the Field Monitoring System for Agriculture, where the water system is done just when there is an extraordinary requirement for water as shown by soil humidity.

Implanted frameworks are the product utilized in the plan. An inherent framework is a PC equipment and programming mix that is either programmable or fit for planning a specific capacity or potentially works in a bigger framework. It is a continuing dedicated OS (RTOS)-modified and limited system that often has persistent device limits. In a larger mechanical and electrical device. Water system readiness is practiced by investigation of soil and water level with pressure meters under trickle water system through the mechanization controller gadget in sandy soil in certain water system frameworks. It is critical to keep the material in the field for the maker. This is the advancement of a trickle water system gadget dependent on the Micro-controller, which is an information control framework progressively for increasingly successful checking and the guideline of all dribble water system activities.

The Horticultural Field Monitor system enables the rancher to use electronic frame methods to apply the right water volume at the right time irrespective of work available for rotating valves. In order to establish a working water system framework, numerous water system frameworks are used in various yields. The humidity and temperature of the plants are directed accurately with this product. Because of the changing conditions, the situation in large farmhouses may in some cases contrast, making it amazingly hard to physically set up consistency at each segment of the farmhouse. GSM is utilized, therefore, to report water system data. The GSM connection is sent to the mobile device. The mix of programming and equipment gives an exceptionally complex impact over the manual procedure. This requires the utilization of the system for remote checking and dribble water system control. Sensors including moistness, soil humidity, and raindrop are utilized. Submit microcontroller qualities to these sensors. Utilizing sequential network, the microcontroller sends data to the PC. The consistent graph estimations of real-time sensors are demonstrated utilizing the Web and Android programming on the PC- and IoS-based mobile. On the off chance that the sensor is arriving at the edge level, the dribble water system segments will naturally be worked by the microcontroller, if the limit data is held. Additionally, through Android mobiles, clients can follow drip irrigation from anyplace.

This is a continuous criticism framework for the checking and guideline of all dribble water system framework tasks in the microcontroller-based trickle water system process. The water system framework oversees valves by methods for robotized ON/OFF controllers which, paying little mind to whether the activity is important to switch the valves or ON/ OFF engine, empowers the rancher to include the perfect sum at the right time. This reduces precipitation on sullied fields, wiping out flooding at improper hours of the day. It builds ranch effectiveness and causes in all regards to set aside cash. The administration of such ranches in any nursery includes an assortment of data and movement to a control unit that is typically situated in a control room expelled from the generation region. Today, a proper link correspondence framework, for example, a field transport, permits the exchange of information from the nurseries to the control framework. Regardless of whether it appears to be very engaging substitute of the wired connection with a remote framework, a completely remote framework may have certain issues. The framework shows the structure of a temperature and soil moistness sensor that can be situated where the temperature and humidity of the soil can be estimated on a field that is touchy to two parameters.

The parameters are subject to air temperature, ground humidity, contamination, and precipitation. The measure of water accessible for the water system was then decided, and the pertinent impacts are determined by the fitting technique, natural components, evapotranspiration, and type of plant. The GSM is one of the most dependable and effectively available remote correspondence systems. The nature of its handset unit and its administrations' membership expense is little, so it is additionally very financially savvy. The inserted gadget interfaced with the GSM module will expand the range of implanted format and inspire the product cluster of the directing and checking frameworks in huge outside valves or engines ON/OFF. This improves plant quality and figures out how to spare vitality. The administration of this kind of ranches includes gathering and transmitting data into a control unit, typically in a control room, aside from the generation territory, in every nursery. Today, a suitable wired correspondence framework, similar to a field transport, ensures that the transference of data among nurseries and the control framework.

In spite of the fact that the substitution of wired frameworks for totally remote frameworks can be engaging, a completely remote framework may have a scope of disadvantages. Most frameworks prescribe that a temperature and ground humidity sensor can be introduced in suitable field areas to control soil temperature and moistness, both of the parameters to which the harvests are subject. In such cases, this sensor is engaged progressively, given the momentary temperature and humidity data, on a criticism control system with a brought-together control unit that controls water stream to the field. The information factors were determined, for example, air temperature, soil moistness, radiation, and humidity. The measure of water fundamental for water system has likewise been assessed, and related outcomes mimicked utilizing the proper strategy, biological conditions, evapotranspiration, and harvest type.

A GSM-based imaginative installed remote control framework for the water system is proposed for this part. On the off chance that the user is inside the 10-m extent of the structured framework, the arrangement and collaboration between the customer and the machine is carried out by means of SMS on the GSM system. Pakistan is the establishment of the global economy and cultivating. The focal point of cultivating is the water system. During adequate precipitation time, the water system is utilized to continue developing yields in the area. About 20% of ranchers in India rely upon electric siphons for the water system in their fields. There are a ton of waste issues.

- Farmers' farmland is commonly far away from the cultivating house, so they need to utilize water system land that causes disadvantages and fuel utilization (if any vehicle is utilized).
- Farmers will have the option to realize the vitality condition of the homestead as the power supply presence is very impermissible.
- Engine consuming happens regularly because of startling voltage varieties and dry riding.
- The presence of malicious part in the agribusiness area is likely.
- Agricultural ranchers are dynamic as pesticides are showered. The prosperity of cultivators is averse to these pesticides.

In the proposed framework, every one of these issues is talked about. The mobile will hand off vitality to the client by means of the GSM organize through Bluetooth/SMS. The gadget controls the siphon's water stream. The framework will hand off client data through GSM organize instant messages when vitality is accessible; however, there is no water. Temperature and humidity sensors are situated nearby and hand off the data to users and soil humidity and surrounding temperature control. The customer sends information to GSM system as SMS to start or stop the water system, as required. With the assistance of the Wi-Fi module, the sensor data are put away on a server. Such qualities are utilized to assess and follow the gadget more.

3.3   ISSUE STATEMENT

The water system of plants ordinarily takes quite a while; much HR should have been practiced in a sensible measure of time. Individuals have generally finished these measures. Numerous offices presently use programming to decrease the number of workers or the time required for watering the harvests. Power is negligible with these gadgets, and different resources stay wasteful. Water is one of the profoundly abused resources. One procedure of watering the yield is the mass water system. This methodology is an immense misfortune because the measure of water provided addresses with the issues of yields. Abundance water is evacuated in nurseries through the gaps of the compartments or permeates through the fields through the earth. The ebb and flow impression of water is that of an available open, sustainable power source ware. This is not valid, be that as it may; water utilization is exhausted in numerous pieces of North America. So it is reasonable to expect that it will immediately turn into a costly item in any piece of the world. Just as the abundance water costs, work turns out to be likewise costly. As an outcome, there will be more cash associated with a similar framework if there is no endeavor to amplify such resources. Programming is undoubtedly a cost decrease technique to diminishing vitality consumption.

3.4   PLAN AND IMPLEMENTATION

3.4.1   BLOCK DIAGRAM

The block diagram of the Agriculture field monitoring system was shown in Figure 3.1. The contemporary necessity to improve plant development and decrease costs underpins the structure of a coordinated water system framework to moderate contamination and expanding remaining tasks at hand and overhead controls. Input-based arrangements make for a more viable asset the board than open-circle models to the cost of trouble and unwavering quality issues. Soil mugginess is hard to gauge, and its objective sums cannot be dependably protected. For a local location, a model is recommended. It comprises durable materials and costs generally low. The various segments were displayed and assessed and exhibited their presentation in water utilization and human mediation decrease. The model is additionally successful all alone with low-power utilization. In any case, a ton of research must be completed in the general procedure to decide the real reserve funds in water and labor. The arrangement contains two sections: the transmitter and the recipient. There will be the entire telecaster portion on the property, which is a versatile and web server as the recipient segment.

At Transmitter

- Arduino Uno
- Moisture sensors
- Temperature sensor
- Raindrop sensor
- LCD display
- GSM module
- Wi-Fi module
- Buzzer
- Power supply
- Water flow system

At Receiver

- Mobile phone
- Web server

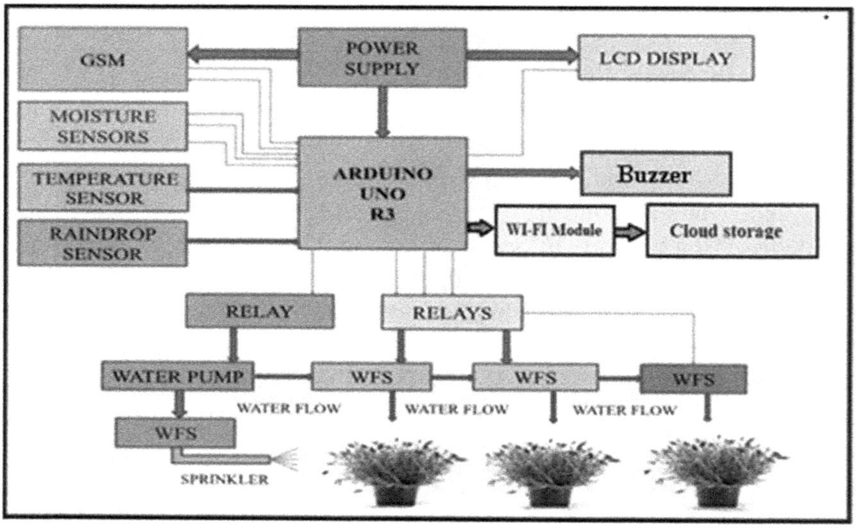

FIGURE 3.1   Block diagram of agriculture field monitoring system.

As appeared in the square graph, the transmitter partition includes various segments. The Arduino UNO unit, viewed as the core of the endeavor, is the focal part for our plan. The key modules are the sensors receiving Arduino; we utilize three humidity sensors, two raindrop sensors, and a sensor for temperature. For any district, a temperature sensor is suitable, while the moistness and the raindrop sensor vary in like manner relying upon the zone. On the off chance that the region is vast, at that point, higher numbers of humidity sensors are required, and if the region is low, just a couple of humidity sensors are utilized.

The clarification is that the humidity levels are not the equivalent all through the area. The plan is additionally an LED screen that shows the qualities controlled by the sensors and utilizes a ringer to flag those working at the field. There, we utilize a submerged gadget, to which a transfer is connected, it is a system that is electrically controlled, near the tapping procedure. The channels are additionally used to give the plants heat. The GSM modem is one of our key units. This GSM modem is utilized to send and get the rancher's information.

The GSM modem just goes about as a mobile. The SIM card is utilized to send and get information from the client. The Wi-Fi gadget that is utilized to store the qualities entered on a web server is another significant unit. For Arduino, humidity, downpour fall indicator, LCD and Wi-Fi unit, and 12 volts of supply are required for the GSM modem, temperature sensor, and transfer capacities, 5 V is an inventory of vitality.

3.4.1.1   WORKING

The sensors of mugginess are sent in different territories of the field and downpour, and temperature sensors are put starting from the earliest stage a specific height. The humidity sensor screens the moistness substance of the soil. The locator of precipitation detects the precipitation. The temp indicator identifies the nature of the earth. When the machine is turned on, the movement of the engine is controlled. If the generator has fuel, it tests for humidity, and generally an estimation of the content is sent to the rancher saying that "the homestead needs water yet the motor isn't running."

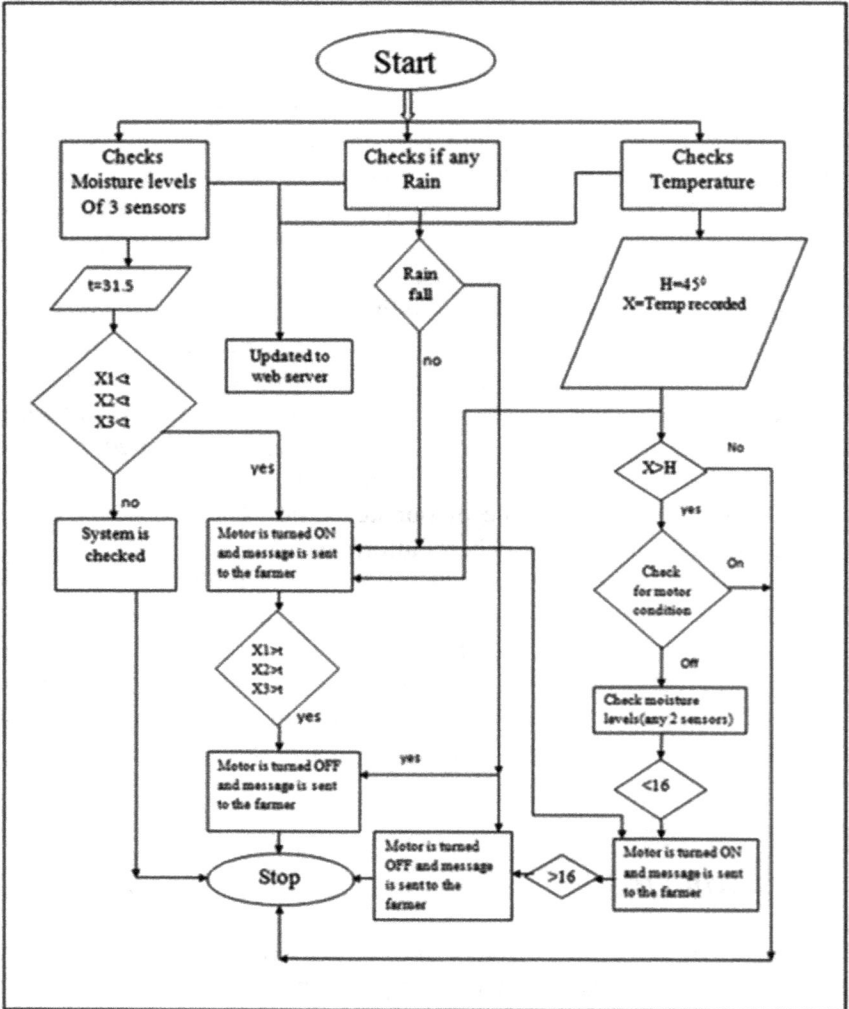

FIGURE 3.2   Agricultural field surveillance system flow chart.

Content "OFF-turned motor" is sent using GSM to the rancher, and on LCD, ringer rings. If during the water system cycle, the precipitation occurs, the downpour drop finder identifies the recurrence of the precipitation isolated in lower precipitation. At the point when a substantial downpour occurs, the generator is turned on by a hand-off. "The engine started because of overwhelming precipitation" is a message sent on LCDs and the bell rings to the rancher.

The exploration has been totally done on the ground up to now. They ought to center around the surface, yet additionally, deal with the leaves during summer or high temperatures. When there is extreme heat, the leaves are occasionally dry. A temperature sensor intermittently identifies the temperature and sends the sign to Arduino if the temperature is higher than the predefined data, for example, 45°C. Through method for a circle, the Arduino turns on the motor. The content "Expanded high-temperature generator" is sent by means of a GSM unit to the rancher, which is demonstrated all the while on LCD and bell rings. The Wi-Fi unit for further information investigation and sensor following changes a humidity sensor data, a raindrop sensor data and a temperature sensor data on the web server.

3.4.1.2   PIN DIAGRAM

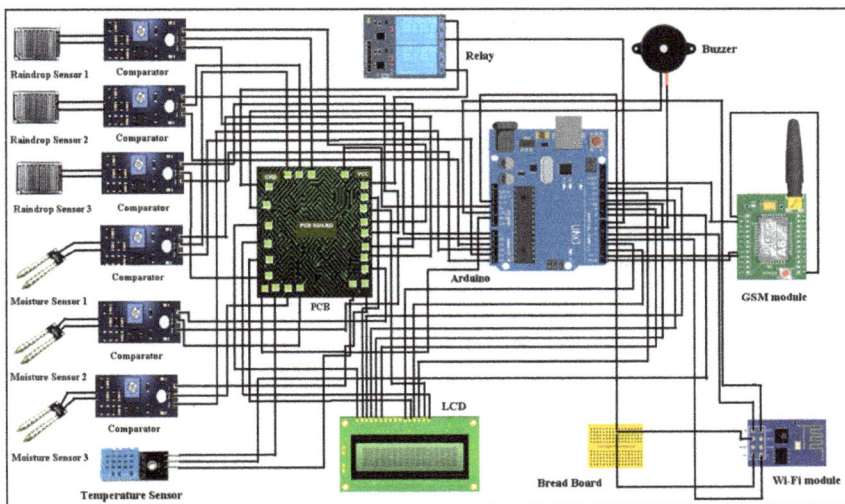

FIGURE 3.3   Pin diagram of agriculture field monitoring system.

- Arduino is the main microcontroller that consists of 14 digital I/O pins (0–13) and 6 analog input pins (A0–A5).
- Analog connections of Arduino:

  A0 is connected to Raindrop sensor-1.

  A1 is connected to Moisture sensor-1.

  A2 is connected to Raindrop sensor-2.

  A3 is connected to Moisture sensor-2.

  A4 is connected to Moisture sensor-3.

  A5 is connected to Raindrop sensor-3.
- Digital connections of Arduino:

  0 is connected to GSM Tx pin.

  1 is connected to GSM Rx pin.

  2 is connected to LCD pin 4.

  3 is connected to LCD pin 6.

  4 is connected to Wi-Fi Tx pin.

  5 is connected to Wi-Fi Rx pin.

  6 is connected to Buzzer.

  7 is connected to Relay.

  8 is connected to Temperature sensor.

  9 is connected to LCD pin 11.

  10 is connected to LCD pin 12.

  11 is connected to LCD pin 13.

  12 is connected to LCD pin 3.

  13 is connected to LCD pin 14.
- Power connections of Arduino :

  5 V is connected to PCB board.

  GND is connected to PCB board.

  3.3 V is connected to Breadboard.
- The 3.3 V from breadboard is connected to Wi-Fi power and channel reset pins.

The PCB board is connected to each VCC and GND pin module

3.4.2   HARDWARE AND SOFTWARE

3.4.2.1   ARDUINO

Arduino is a prototyping apparatus for open-source innovation to make the way toward utilizing gadgets progressively accessible for multidisciplinary ventures. The framework is a fundamental open improvement of equipment for the standard Arduino programming language board and the boot loader working on the machine.

Through accepting a contribution from different sensors, Arduino can detect the climate and can screen lights, engines, and other drive frameworks in its condition. The framework microcontroller is planned with the programming language of Arduino (base on wiring) and Arduino (in light of handling).

Arduino assignments can be self-governing or work on the machine through code. The sheets can be worked by hand or pre-amassed; the product is allowed to utilize. You can utilize open-source consent to adjust the equipment reference structure (CAD documents) to your prerequisites. Arduino equipment is planned with a link-based language like C++. Streamlined, refreshed, and deciphered IDEs, the Arduino equipment is likewise furnished with C++ libraries.

Arduino is a microcontroller that can detect and screen PCs. It is a stage-dependent on an ATmega328 microcontroller board and a PC programming advancement structure. Arduino plans are self-governing and can be run on PC show gadgets. Pre-gathered frameworks are basic or can be worked by our needs, open-source IDE. The Arduino programming language chips away at wires, a comparative physical stage dependent on the sight and sound programming preparing condition.

A typical Arduino board appears as shown in Figure 3.4. The Arduino Uno ATmega328 uses the equivalent microcontroller. The hardware consists of a main calibrated system framework with onboard details and Arduino board distribution assistance. The software consists of a nonexclusive parser of the programming language and the Arduino boot loader.

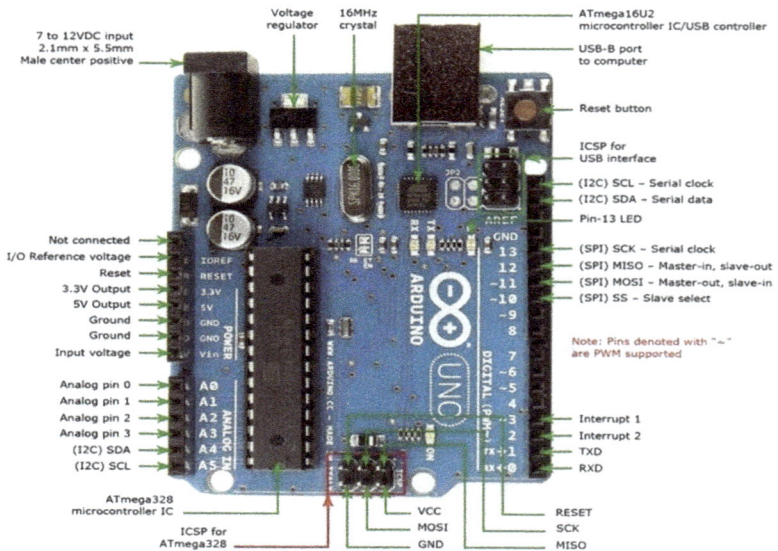

FIGURE 3.4 Arduino board.

3.4.3 FEATURES OF ARDUINO ATMEGA328

- High-performance, Low-power AVR® 8-Bit Microcontroller
- Advanced RISC Architecture
  - 131 Powerful Instructions – Most Single Clock Cycle Execution
  - 32 × 8 General Purpose Working Registers
  - Fully Static Operation
  - Up to 20 MIPS Throughput at 20 MHz
  - On-chip 2-cycle Multiplier
- High Endurance Nonvolatile Memory Segments
  - 32K Bytes of In-System Self-Programmable Flash program memory
  - 1K Bytes EEPROM
  - 2K Bytes Internal SRAM
  - Write/Erase Cycles: 10,000 Flash/100,000 EEPROM
  - Data retention: 20 years at 85°C/100 years at 25°C
  - Optional Boot Code Section with Independent Lock Bits

- In-System Programming by On-chip Boot Program
- True Read-While-Write Operation
- Programming Lock for Software Security
- Peripheral Features
  - Two 8-bit Timer/Counters with Separate Prescaler and Compare Mode
  - One 16-bit Timer/Counter with Separate Prescaler, Compare Mode, and Capture Mode
  - Real-Time Counter with Separate Oscillator
  - Six PWM Channels
  - 8-channel 10-bit ADC in TQFP and QFN/MLF package Temperature measurement
  - 6-channel 10-bit ADC in PDIP Package Temperature Measurement
  - Programmable Serial USART
  - Master/Slave SPI Serial Interface
  - Byte-oriented 2-wire Serial Interface (Philips I2C compatible)
  - Programmable Watchdog Timer with Separate On-chip Oscillator
  - On-chip Analog Comparator
  - Interrupt and Wake-up on Pin Change
- Special Microcontroller Features
  - Power-on Reset and Programmable Brown-out Detection
  - Internal Calibrated Oscillator
  - External and Internal Interrupt Sources
  - Six Sleep Modes: Idle, ADC Noise Reduction, Power-save, Power-down, Standby, and Extended Standby.
- I/O and Packages
  - 23 Programmable I/O Lines
  - 28-pin PDIP, 32-lead TQFP, 28-pad QFN/MLF and 32-pad QFN/MLF
- Operating Voltage:
  - 1.8–5.5 V
- Temperature Range:
  - 40–85°C
- Speed Grade:
  - 0–4 MHz @ 1.8–5.5 V, 0–10 MHz @ 2.7–5.5 V, 0–20 MHz @ 4.5–5.5 V.

- Power Consumption at 1 MHz, 1.8 V, 25°C
  - Active Mode: 0.2 mA
  - Power-down Mode: 0.1 µA
  - Power-save Mode: 0.75 µA (Including 32 kHz RTC)

3.4.4   PIN DIAGRAM ATMEGA328

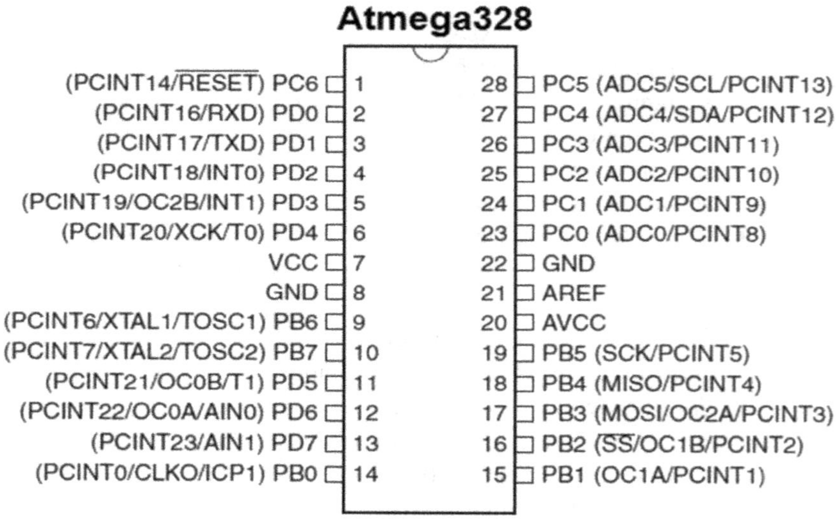

Atmega328

(PCINT14/RESET) PC6 ☐ 1	28 ☐ PC5 (ADC5/SCL/PCINT13)
(PCINT16/RXD) PD0 ☐ 2	27 ☐ PC4 (ADC4/SDA/PCINT12)
(PCINT17/TXD) PD1 ☐ 3	26 ☐ PC3 (ADC3/PCINT11)
(PCINT18/INT0) PD2 ☐ 4	25 ☐ PC2 (ADC2/PCINT10)
(PCINT19/OC2B/INT1) PD3 ☐ 5	24 ☐ PC1 (ADC1/PCINT9)
(PCINT20/XCK/T0) PD4 ☐ 6	23 ☐ PC0 (ADC0/PCINT8)
VCC ☐ 7	22 ☐ GND
GND ☐ 8	21 ☐ AREF
(PCINT6/XTAL1/TOSC1) PB6 ☐ 9	20 ☐ AVCC
(PCINT7/XTAL2/TOSC2) PB7 ☐ 10	19 ☐ PB5 (SCK/PCINT5)
(PCINT21/OC0B/T1) PD5 ☐ 11	18 ☐ PB4 (MISO/PCINT4)
(PCINT22/OC0A/AIN0) PD6 ☐ 12	17 ☐ PB3 (MOSI/OC2A/PCINT3)
(PCINT23/AIN1) PD7 ☐ 13	16 ☐ PB2 (SS/OC1B/PCINT2)
(PCINT0/CLKO/ICP1) PB0 ☐ 14	15 ☐ PB1 (OC1A/PCINT1)

FIGURE 3.5   Pin diagram of Atmega328.

Arduino is the key microcontroller that drives the entire circuit in this structure. Through killing and the hand-off, Arduino distinguishes all sensor esteems as shown in Figure 3.5.

3.4.4.1   GSM MODEM

GSM is a SIM-card-enabled gadget that runs through a portable admin, much like the mobile one as shown in Figure 3.6. From the perspective of the portable manager, the GSM modem functions as a mobile. It requires the gadget to utilize a GSM framework to chat on a versatile system when a GSM transmitter is joined to a machine. Although these GSM modems are often used to provide mobile web access, most can also be used to warn SMS and MMS. SMS Lite will only send and receive SMS and MMS messages via GSM modem.

A GSM modem could be a committed modem unit, USB, Bluetooth, or a mobile modem with limitations to the GSM modem. An email can be sent and prepared by the portable administrator as though it was really sent on mobile. A GSM modem will acknowledge an extended AT control assortment for sending and accepting instant messages all together for these capacities to be performed.

FIGURE 3.6   GSM modem.

Specifications:
Uplink frequency: 876–915 MHz
Downlink frequency: 921–960 MHz
Channel number: 955–1023
Working voltage: 4.5–5.5 V DC
Working current: max of 2 A
Baud rate: 115,200 bps

GSM modems may function as an efficient and fast means of using SMS since no individual member of an SMS professional association is needed. GSM modems are a practical solution in many parts of the world to accept SMS messages as the recipient charges for the transmission. An Ethernet, Wi-Fi, or LAN interface could be the GSM modem. A GSM modem may also be a regular GSM phone attached to a sequential port or a USB port on a gadget that has the right link and application manager.

At commands

AT	Check if serial interface and GSM modem is working.
ATED	Turn echoes off, less traffic on the serial line.
AT+CNMI	Display of new incoming SMS.
AT+CPMS	Selection of SMS memory.
AT+CMGF	SMS string format, how they are compressed.
AT+CMGR	Read new message from a given memory location.
AT+CMGS	Send a message to a given recipient.
AT+CMGD	Delete the message.

GSM is used in this work for sending messages from Arduino to the producer, including sensor values and engine conditions.

3.4.4.2   WI-FI MODULE

The ESP8266 is a cost-effective Wi-Fi microchip that comprises a complete TCP/IP stack and an Espressif Systems producer microcontroller in Shanghai, China, as shown in Figure 3.7.

FIGURE 3.7   Wi-Fi module.

Specifications:
Processor: 32 bit RISC
Frequency: 80 MHz

Vcc: 3.3–3.6 V
Memory: 32 KB

For August 2014, the ESP-01 model developed by the external whole-saler Ai-Thinker was taken into account by the Western manufacturers. A tiny module helps microcontrollers to communicate with Wi-Fi and to make basic TCP/IP connections with the Hayes-style instructions. From the beginning, however, it had almost no English documentation on the chip and its instruction. The ESP8285 is a 1 MiB streamed, ESP8266 that can interface with Wi-Fi by chip-only. The ESP32 will suit the microcontroller chips released in 2016. ESP8266 is a Wi-Fi-enabled chip (SoC) framework produced by the Espressif framework. This is used for the structure of IoT installed frameworks.

At Commands

AT+CWMODE	To activate WI-FI mode
AT+CWJAP	To join AP
AT+CWLAP	To list AP
AT+CIPSEND	To send the data

The Wi-Fi unit in this chapter is used to upload the sensor values to the web server so that device tracking from any position can be performed.

3.4.4.3 SENSORS

Sensors are specific instruments regularly utilized for the recognizable proof and reaction of the electrical or optical sign.

3.4.4.3.1 Humidity Sensor

Sensors of soil moistness compute the groundwater content. A soil mugginess sensor comprises a few humidity indicators for the soil. Recurrence area locator, for example, a capacitance sensor, is generally utilized for soil humidity sensors.

- Moisture gage neutron, utilizing neutron fluid mediator property.
- Soil electrical quality.

We will utilize the humidity sensors that can be embedded into the soil to decide the soil's mugginess content in this specific section. The electric conductivity of the soil is resolved absolutely by utilizing two isolated metal conductive, with the particular case that the broke up salts change water conductivity significantly and can convolute figuring. One modest fix is to fuse conductors into a permeable square of gypsum that discharges calcium and sulfate particles with the goal for them to overwhelm the ground level of particles. A tertiary pointer for use in medium and enormous soils is associated with the flood-assimilated water possibilities over the range 60–600 kPa. Non-dissolving granular network sensors for the range 0 to 200 kPa are currently usable, using inside estimation methods to alleviate inconstancy because of solutes and temperature. Intermediary factors that structure a component by method for soil electrical conductance and are hence characteristically inclined to changes in soil saltiness, temperature, and water are estimated by techniques for the handling of dielectric soil properties. The surface mass volume and the bound and free water proportion of soil type regularly impact estimations.

In unique conditions, sensible affectability and accuracy can be gotten, and some sensor structures for logical work have been broadly received.

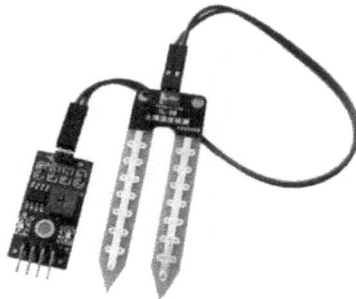

FIGURE 3.8   Moisture sensor.

Specifications:
Input voltage: 3.3–5 V
Output voltage: 0–4.2 V
Input current: 35 mA
Output signal: Both analog and digital

As a rule, changes to volumetric humidity substance and fluid limit by auxiliary or tertiary strategies from crude sensor readings seem, by all accounts, to be sensor- or soil-explicit, at elevated levels of saltiness and temperature subordinate. The accuracy estimated in lab estimations for examining grade instruments is commonly more regrettable than ±4% if the production line settings are utilized or not exactly ±1% if the soil is adjusted. TDR-based sensors additionally will, in general, need less tuning, yet cannot be custom-fitted to high-salt and earth containing soils. For granular network sensors, there is no practically identical research center determination, potentially because they are harder to align, their reaction times are generally moderate, and the yield is hysteric for wetting and drying bends. Dielectric soil estimation is the special technique for most research considers where testing, execution, and examination are conceivable; however, because of the plausibility of stray capacitance, the potential for cost decrease by sensor multiplexing is restricted. The structure of enhanced Application Specific Circuits (ASICS), however, needs a considerable degree of the venture and would decrease producing costs. Different sensors must give a profundity profile and an agent region; however, these expenses can be limited by utilizing a PC model to extend estimations presciently. Hence, using humidity sensors, the abrogating factor for information access and examination is precise, financially savvy sensors and electronic frameworks.

The humidity sensor is utilized in this section for the estimation of the measure of humidity on the soil subject to opposition as shown in Figure 3.8.

3.4.4.3.2 Temperature Sensor

The DHT11 is a customarily utilized sensor for temperature and mugginess. The indicator accompanies a different NTC for temperature estimation and an eight-piece microcontroller for sequential data to create temperature and moistness data. The sensor is additionally easy to understand, so it is simple for different microcontrollers to the interface.

Sensors calculate temperatures from 0 to 50°C for accuracy of ±1°C and ±1% for the humidity of 20–90%. The DHT11 sensor is designed to provide sequential data, which makes it easy to adjust. The moisture detection component of the DHT11 is moisture support with superficial cathodes. The partial particles are released via the substratum when the surface gets water fume, which increases the conductivity between the anodes.

The disparity between the two interfaces is proportional to the overall humidity. Higher relative moisture reduces the difference between terminals when lower relative moisture increases obstruction of the anode. The DHT11 adjusts the calculation for relative moisture safety at the back of the device chip and transmits moisture readings to the Arduino Uno.

FIGURE 3.9   Temperature sensor.

Specifications:
Operating voltage: 3–5 V
Temperature range: 0–50°C
Humidity range: 20–80%
Max current during measuring: 25 mA

The temperature sensor tests the humidity and temperature of the ambient area with the aid of the thermistor within the device as shown in Figure 3.9.

3.4.4.3.3   Raindrop Sensor

A climate locator or an atmosphere pointer is the precipitation change. Two primary downpour sensor applications are accessible. The first is a robotized water system gadget that enables the procedure to close down if there is a downpour. The second is a gadget used to ensure the vehicle's inside against rain and bolster windscreen consequently. Furthermore, a downpour blower will be activated in the receiving wire feed opening in proficient satellite correspondences reception apparatuses, to dispose of water beads from the Mylar spread, which keeps up weight and dry air in the wave direction frameworks.

Remotes just as designed, downpour sensor for water system frameworks, are conceivable, most of which are hygroscopic circles that swell within the site of a downpour to diminish again while the electric switch is kept dry and discharged by the hygroscopic plate stack. Numerous instruments, however, are additionally sold using tipping cans or lead-style indicators to screen precipitation. Wired and remote models likewise perform related components to suspend water system controllers briefly. These are either connected to the sensor terminals of the water system controller or are set on the standard circuit of the solenoid valve arrangement to maintain a strategic distance from the opening of valves in the wake of identifying water.

Some irrigable downpour sensors additionally have a stop sensor to shield the framework from solidifying at frigid temperatures, mainly when water system frameworks keep on being utilized throughout the winter.

FIGURE 3.10   Raindrop sensor with module.

Specifications:
5 cm × 4 cm nickel plate
Driving ability over 15 mA
Working voltage: 5 V

The rain sensor module is an easy tool for rain detection as shown in Figure 3.10. It can be used as a switch when raindrop falls through the raining board and also for measuring rainfall intensity. The module features, a rain board and the control board that is separate for more convenience, power indicator LED and an adjustable sensitivity though a potentiometer. The analog output is used in detection of drops in the amount of rainfall. Connected to 5V power supply, the LED will turn on when induction board has no rain drop, and DO output is high. When dropping a little amount water, DO output is low, the switch indicator will turn on. Brush off the water droplets, and when restored to the initial state, outputs high level.

3.4.5   LCD DISPLAY

The name is itself "Fluid Crystal" shown in Figure 3.11. It is, without a doubt, a blend of the strong and the wet condition of the material. These both have the highlights of solids and fluids and hold their separate states. The reasonable preferred position of the fluid gem show is that it has a lower control yield than the LCD. It is generally the microwave arrangement of the light contradicted to a specific factory watt succession of LEDs. This

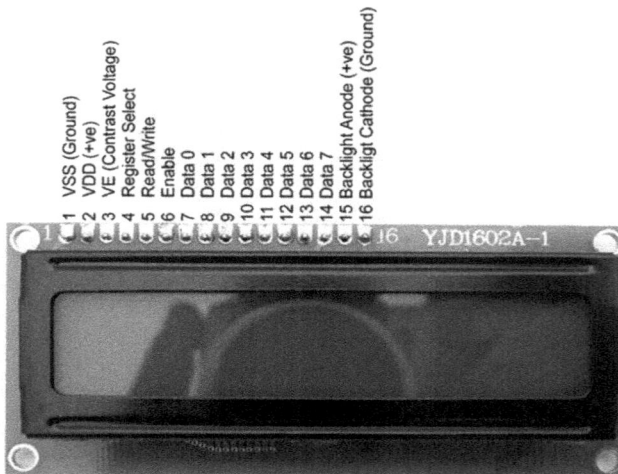

FIGURE 3.11   Liquid-crystal display.

is consistent with MOS worked in rationale circuit through low-power utilization particulars. Different favorable circumstances incorporate their low-cost and substantial examination.

3.4.6 LIQUID CRYSTAL DISPLAY

Specifications:

Dimension: 16×2
Supply voltage: 5 V
Data pins: 8

A thin layer (around 10 μm) of the fluid precious stone is shaped by a fluid gem cell sandwiched between two glass sheets with the straightforward electrical tubing on their internal appearances. The cell is known as a transmitting sort of cell when both glass sheets are straightforward. The cell is alluded to as the intelligent structure since one glass is translucent and different and has an intelligent covering. The LCD gives no light of its own. In all actuality, it relies upon the enlightenment from an outer hotspot for its specific visualization. The LCD is utilized to show the absolute number of characters, up to 16 characters of every 2 lines in this 16×2 dimensional plan.

3.4.7 TRANSFER

A transfer is an association of power. Most transfers work precisely utilizing an electromagnet. However, other operational ideas are additionally used. Unwinds are utilized when the lower control signal is required for controlling the circuit or when there is one sign to work different circuits. Unwinding has been broadly utilized for down-to-earth forms in portable trades and early PCs.

As a contractor, there is a sort of transfer that can handle the high power required for the direct control of the motor or various burdens. Solid state switches regulate circuits that have no design sections instead of using a semiconductor. These activities were augmented by electronic gadgets dubbed "protection transfers," some of which are used to defend electric circuits from overburden and loss of modern electrical power frameworks as shown in Figure 3.12.

FIGURE 3.12   Relay.

Specifications:

Supply voltage: 5 V DC
Supply current: 70 mA
Operating time: 10 ms
Maximum switching: 300 operations/min

When an electric flow crosses the loop, an attractive field is made, which starts at the edge, and the subsequent development of a moving contact is the moment of truth, the association with permanent contact. When the hand-off is de-stimulated, the process opens up connections and dismantles the connection, when the amount of contacts is closed. The weight returns the armature to its quiet area, about a large part of the strength of the magnet potency, when the current to the circuit is murdered. It is normally supplied by a spring, but inertia is also commonly used in modern engine starters. Most transfers are quickly manufactured. When used in low-voltage applications, it has an impedance; when used in high voltage or current, this decreases arcing.

At the point where the loop is powered by direct power, an auto-transfer diode is regularly positioned over the curl to disperse the vitality from the attractive falling field at deactivation. The flood may likewise be consumed by touch security system made up of a succession resistor and condenser (snubber circuit).

A little espresso concealing ring can be incinerated as far as possible of the solenoid, which delivers a slight out-of-stage current, which improves minor weight in the section during the AC procedure. At the point when the loop is worked to be empowered by rotating power (AC). Hand-off goes about as a switch in this venture, killing the motor and as per the sensor data and the Arduino answer.

3.4.8 SUBMERSIBLE PUMP

A submersible siphon (ESP) is a mechanical system shown in Figure 3.13 that has an airtight encased motor close to the cylinder part. The whole establishment is isolated into the siphoning liquid. It evades the siphon cavitation as an issue identified with an enormous hole in raising between the cylinder and the fluid surface. Rather than fly siphons that must force liquids, submersible siphons push the liquid to the surface. Submersibles are more secure than cylinder motors.

Multi-stage radial siphons in a vertical position are the submersible siphons utilized in ESP frameworks. In spite of the fact that its development and methodology have changed throughout the years, the basic guideline of administration continues as before. Created fluids lose their motor vitality when they experience enormous outward powers because of high impeller rotational speed when they are changed over to pressure vitality in the diffuser. This is the essential working strategy for blended and outspread stream siphons

FIGURE 3.13 Submersible pump.

The pump shaft is connected to the gas separator or the protector by a mechanical coupling at the bottom of the pump. When fluids enter the pump through an intake screen and are lifted by the pump stages. Other parts include the radial bearings (bushings) distributed along the length of the shaft providing radial support to the pump shaft turning at high rotational speeds. An optional thrust bearing takes up part of the axial forces arising in the pump but most of those forces are absorbed by the protector's thrust bearing.

3.4.9 MOBILE

Mobile, as depicted in Figure 3.14, is a framework that can make phone calls through radio associations and goes across enormous geological territories, frequently known as mobile phone, wireless, and hand devices. This can be accomplished through associations with a portable system gave to the general population phone systems by a mobile administrator. A link broadcast communications, however, is just utilized in a solitary, individual base station inside the short separation. A wide scope of different highlights, including instant message, MMS, fax, Internet access, infrared, Bluetooth, business applications, and games, just as photography is empowered by the state-of-the-art mobiles.

FIGURE 3.14   Mobile phone.

The mobile is a collector for this venture to gather and show messages sent by means of the GSM unit in the inbox.

3.4.9.1 ARDUINO SOFTWARE IDE

The Arduino IDE is the cross-plate design framework of Java and arises from the GUI for the language of coding and ties as shown in Figure 3.15. This aims to bring innovations to veterans and different beginners together. It includes a code editor with usefulness such as punctuation, co-ordination, and programmed indenting, and it also allows programming to be ticked and collected to the board. Typically, there is no compelling reason to change records or execute order line programs.

In the advancement condition of Arduino, a word processor for code composition, a message area, book support, a standard toolbar, and various menus are provided. The Arduino hardware interfaces to access and communicate with applications. Portrayals are called programming composed of Arduino. These drawings are written in the editorial manager of the content. It has examples of material cut/glue and looks/supplant. The message field is entered when you save and send, and errors also arise. The reassurance shows that Arduino content yields complete messages and other data.

Arduino IDE is provided with a C/C++ library known as "wiring" which improves numerous basic I/O tasks. Arduino is written in C/C++, so clients need just set two capacities to build an executable code. For the Arduino Integrated Development Environment (AIDE) or Arduino App, you should recall a content manager for application creation, the message area, the book display, the toolbar with Common Functions grab and a set of menus. It attaches to the Arduino and Genuino PCs to control software and collaborate. Arduino IDE atmosphere is shown in Figure 3.15.

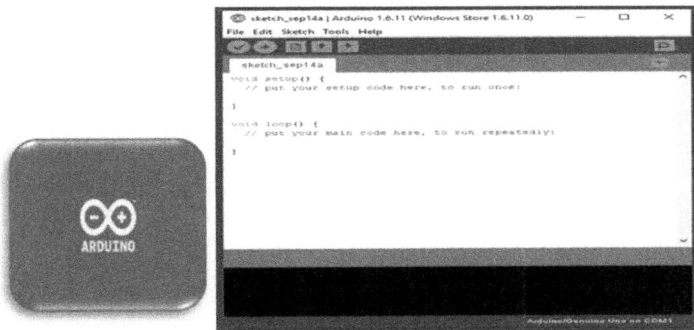

FIGURE 3.15 Arduino IDE environment.

3.4.9.2   COMPOSING SKETCHES

Programs created with Arduino Interface (IDE) are named representatives. Such descriptions were published in the project developer and are not included in the introduction of the .ino folder. The manager has the usefulness of content cut/past and search/supply. The output is inserted in the message area when saving, sending out, and displaying errors. The commodity displays the execution of Arduino Software (IDE), including complete error messages and other data. The monitor and the serial port are on the bottom right of the display. You can track and switch programs, build, open, and delete contours, and use the toolbar to access the sequential window.

Verify
Checks your code for errors compiling it.

Upload
Compile and upload your code to the board configured. For details, see the upload below.

Note: If you use this icon with an external programmer on your board, you can hold a "shift" key on your computer. The message is updated to "Upload programmer."

New
Creates a new sketch.

Open In your sketchbook, you will show a list of all drawings. If you press one, the contents are overridden by the current window.

Note: this menu is not scrolling due to an error within Java, so use the File Sketchbook menu if you need to access a sketch late in the chart.

Save
Saves your sketch.

Serial Monitor
Opens the serial monitor.

More commands in the five menus are found: **File, Edit, Sketch, Tools, Help**. The menus are context-sensitive, which ensures that the work being carried out is visible only in those products. **FILE**

- *New*
  Creates a new instance of the editor, with the bare minimum structure of a sketch already in place.

- *Open*
  Allows to load a sketch file browsing through the computer drives and folders.
- *Open Recent*
  Provides a shortlist of the most recent sketches, ready to be opened.
- *Close*
  Closes the instance of the Arduino Software from which it is clicked.
- *Save*
  Saves the sketch with the current name. If the file has not been named before, a name will be provided in a "Save as.." window.
- *Save as.*
  Allows saving the current sketch with a different name.
- *Preferences*
  Opens the Preferences window where some settings of the IDE may be customized, as the language of the IDE interface.
- *Quit*
  Closes all IDE windows. The same sketches open when Quit was chosen will be automatically reopened the next time you start the IDE

3.4.9.3   EDIT

- *Undo/Redo*
  Goes back of one or more steps you did while editing; when you go back, you may go forward with Redo.
- *Cut*
  Removes the selected text from the editor and places it into the clipboard.
- *Copy*
  Duplicates the selected text in the editor and places it into the clipboard.
- *Paste*
  Puts the contents of the clipboard at the cursor position, in the editor.
- *Select All*
  Selects and highlights the whole content of the editor.

- *Comment/Uncomment*
  Puts or removes the // comment marker at the beginning of each
  selected line.

3.4.9.4   SKETCH

- *Verify/Compile*
  Checks your sketch for errors compiling it; it will report memory
  usage for code and variables in the console area.
- *Upload*
  Compiles and loads the binary file onto the configured board
  through the configured port.
- *Include Library*
  Adds a library to your project by adding # include data at the beginning of your script. See libraries below for more details. However,
  you can enter the Library Manager from this menu item and load
  new libraries from the files.zip.
  Add File...
  Adds to the sketch a source file (copied from its current position).
  In the sketch window, the new file appears in a new tab. Displays
  can be removed with a click on the small triangle icon on the right
  side of the toolbar, which can be accessed from the tab menu under
  the serial screen.

3.4.9.5   TOOLS

- *Serial Monitor*
  Opens the window for the serial screen and initiates the data
  exchange on a wired port panel.
- *Board*
  Select the board that you're using. See below for descriptions of
  the various boards.
  Port
  The menu includes all the serial devices on your machine (actual
  or virtual). Once you access the top-level software tab, it should
  be updated automatically

3.4.9.6 *HELP*

Various archives with the Arduino Interface (IDE) can be effectively gotten to here. You have a neighborhood access, no web association with getting Started, Guide, this IDE manage, and different records. The papers are a reproduction of the reports locally accessible on the Web and can be connected back to our site.

Find in Reference: This is the only online aspect of the Help menu: in the local version of the Guide, you pick the appropriate page directly for the cursor role or order.

3.5 SKETCHBOOK

The Arduino Interface (IDE) pursues the sketchbook idea: a typical spot for holding the projects (or drawings). You can utilize File > Sketchbook list and the Open catch on the toolbar to open the representations in your sketchbook. At the point when you start running Arduino programming, a sketchbook registry is naturally made. The settings discourse encourages you to see or alter the situation of the book page.

3.6 TABS, MULTIPLE FILES, AND COMPILATION

These enable you to use more than one folder (both shown in your own tab) to handle sketches. These may be normal Arduino code files, C files (.c extension), C++ files (.cpp), or header files (.h). These can also be normal.

3.7 TRANSFERRING

Select the right elements from Tools > Panel and Tools > Ports menus before the drawing is moved. There were defined the accompanying boards. The Mac port is likely a sequential variation of sequential TTY.usbmodem241 or /dev/tty.usbserial-1B1 or tty.usbmodem241 (on a Duemilanove or previous USB board). The Windows Gadget area is likely to be COM1 or COM2 (for a sequence board) or COM4, COM5, or COM7 or higher for a UPS board. This will usually be a sequence of the same type. The sequence is usually the same. It should be used for Linux,/dev /ttyACMx,/dev/ttyUSBx and equivalent. When you have selected an appropriate series port and deck, either press the Import button toolbar or choose a Download item from the list Sketch. Show the reset and start to change Arduino sheets accordingly. Just before you start to move on to more well-decided sheets (before Diecimila) that do not automatically reset, you must press reset. In a lot of discussions, you will see the RX and TX LED strings when the image is published. The Arduino Software (IDE) displays a message or a blunder when your transfer is complete.

To transfer a sketch, use the Arduino bootloader, a small program stacked in your board microcontroller. With no additional equipment, you can pass data. The bootloader works a few moments when the board is resetting, and then a drawing that is last forwarded to the microcontroller starts. The onboard LED (stick 13) will be squashed when the bootloader starts (for example, at the stage where the board resets).

3.8   LIBRARY

Library offers additional use to use in pictures, e.g., equipment work or information control. Select this from the Sketch > Import Library menu to use a library in a sketch. At the highest point of the drawing, this will include at least one explanation and gather the book with your drawing. Since libraries with your drawing are moved to the table, they through the storage it requires. If a sketch does not need a library any more, essentially erase its claims by the top of your code.

The list includes a round-up of repositories. Several libraries are included in the programming of Arduino. Many references or library managers can be downloaded. A compressed document will be saved from a library and used in a free sketch, starting with IDE's 1.0.5 form. See these instructions for the construction of an external collection. There are not many collections in our category

- include<LiquidCrystal.h>
- #include <SoftwareSerial.h>
- #include "DHT.h"

3.9   SEQUENTIAL MONITOR

On Arduino and Genuino sheets (USB or Sequential Boards) sequential data will be displayed. Double information showcases. Choose content and snap-on fastening or switching the passage to forward data. Choose the baud rate from below, which is perfect for the value sent to Serial. begin in your skills. Note, once you link to a sequential display (start the execution of the drawing towards the start), the Arduino or Genuino panel is reset to Windows or Mac or Linux.

3.10   SHEETS

The pickup board has two effects: sets the assembly and drawing mounting parameters (e.g., CPU velocity and baud rate) and sets the bootloader consumption record and layout. In the last few words, some board definitions are various. You will have to check whether you transferred a specific choice effectively before using the boatload. The system assistance, which focused everything on Arduino Software AVR Center (IDE), is included in the following round-up process. The board's manager can offer help for new sheets that are always based on the different nucleus, for instance, Arduino Due, Arduino Zero, Edison, and Galileo in the standard establishment.

Arduino/Genuino Uno

An ATmega328 running at 16 MHz with auto-reset, 6 Analog In, 14 Digital I/O, and 6 PWM

3.11   THINGSPEAK

ThingSpeak is a software and an API that provide an open-source IoT framework which is used to extract and retrieve information from objects that use the HTTP protocol on the Internet or via the Local Area Network as shown in Figure 3.16. Initially propelled as a stage for IoT gadgets by ioBridge in 2010, ThingSpeak has incorporated help from MathWorks, MATLAB advanced register to program, empowering ThingSpeak clients to check and envision the information transferred from MATLAB without purchasing a MATLAB permit from MathWorks. ThingSpeak is firmly identified with MathWorks, Inc.

All ThingSpeak documentation is stored for the documentation website for MathWorks, MATLAB, and permits even enrolled client reports for the ThingSpeak site page to be a legitimate sign-in certifications. The terms and conditions for ThingSpeak.com's activity and protection strategy lie between the chose customer and MathWorks, Inc.

FIGURE 3.16   ThingSpeak environment.

ThingSpeak is an IoT application to monitor cloud information streams that can cause you to integrate, display, and disintegrate. From your gadgets, you can send ThingSpeak information, quickly send vital information and send alarms, for example, using Twitter and Twilio web administrations. The coding in MATLAB is generated and implemented in ThingSpeak for handling, description, and analysis. ThingSpeak makes IoT applications for professionals and scientists by building databases and web programming.

- Configures PCs to transmit ThingSpeak data utilizing the REST API and MQTT. ThingSpeak incorporates the accompanying.
- On-request total information from gadgets and sources gave by outsiders.
- Get live or recorded sensor information moment representations.
- Preprocess your gathered information and break it down with MATLAB incorporated.
- Play out the IoT study on a daily schedule or case premise naturally.
- Conduct and transmit the information using systems of outsiders, for example, Twilio or Facebook.

3.12   RESULTS

STEP 1: Open a new window and write the required code in the Arduino IDE software.

STEP 2: Save the code.

STEP 3: Verify the code and check for errors.

STEP 4: Uploading the code onto Arduino UNO.

STEP 5: Switch ON the power supply then power will be distributed to all the components.

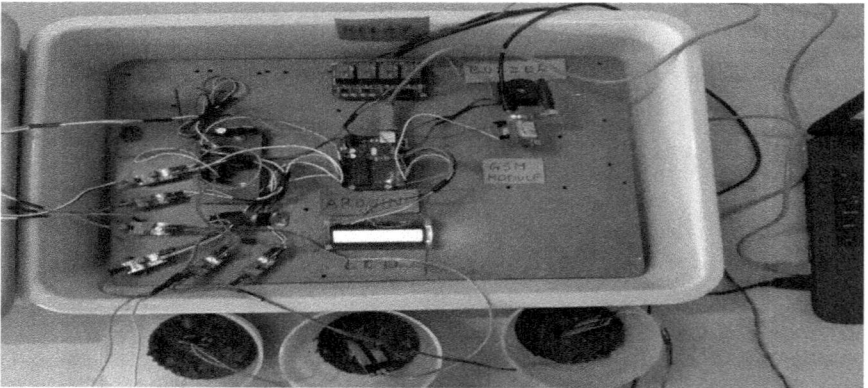

STEP 6: When the soil is dry, then the motor turned on, and the message was sent to the farmer.

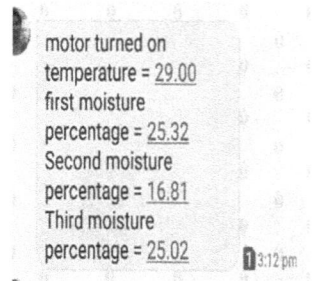

motor turned on
temperature = 29.00
first moisture
percentage = 25.32
Second moisture
percentage = 16.81
Third moisture
percentage = 25.02   3:12 pm

STEP 7: When the soil is wet, then the motor turned off, and the message was sent to the farmer.

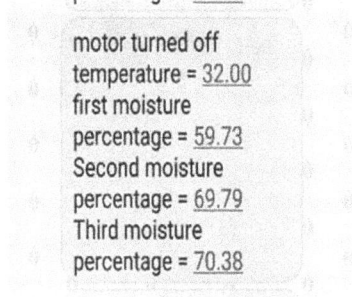

motor turned off
temperature = 32.00
first moisture
percentage = 59.73
Second moisture
percentage = 69.79
Third moisture
percentage = 70.38

STEP 8: When rainfall occurs during motor is on, the motor turned off, and the message was sent to the farmer.

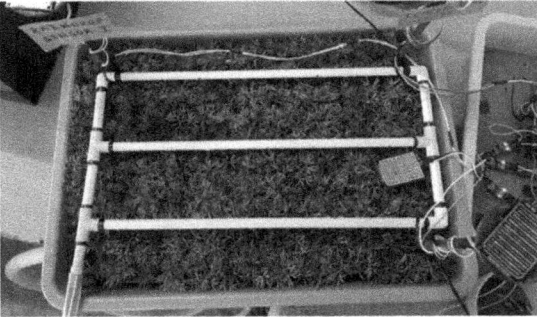

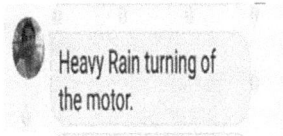

Heavy Rain turning of the motor.

STEP 9: When the temperature crossed 45 degrees, the motor turned on, and the message was sent to the farmer.

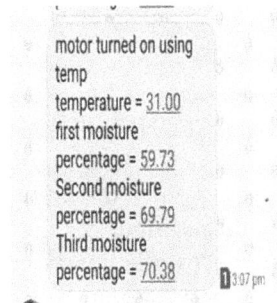

motor turned on using
temp
temperature = 31.00
first moisture
percentage = 59.73
Second moisture
percentage = 69.79
Third moisture
percentage = 70.38   3:07 pm

STEP 10: The recorded values of the moisture sensors, raindrop sensors, and temperature sensor.

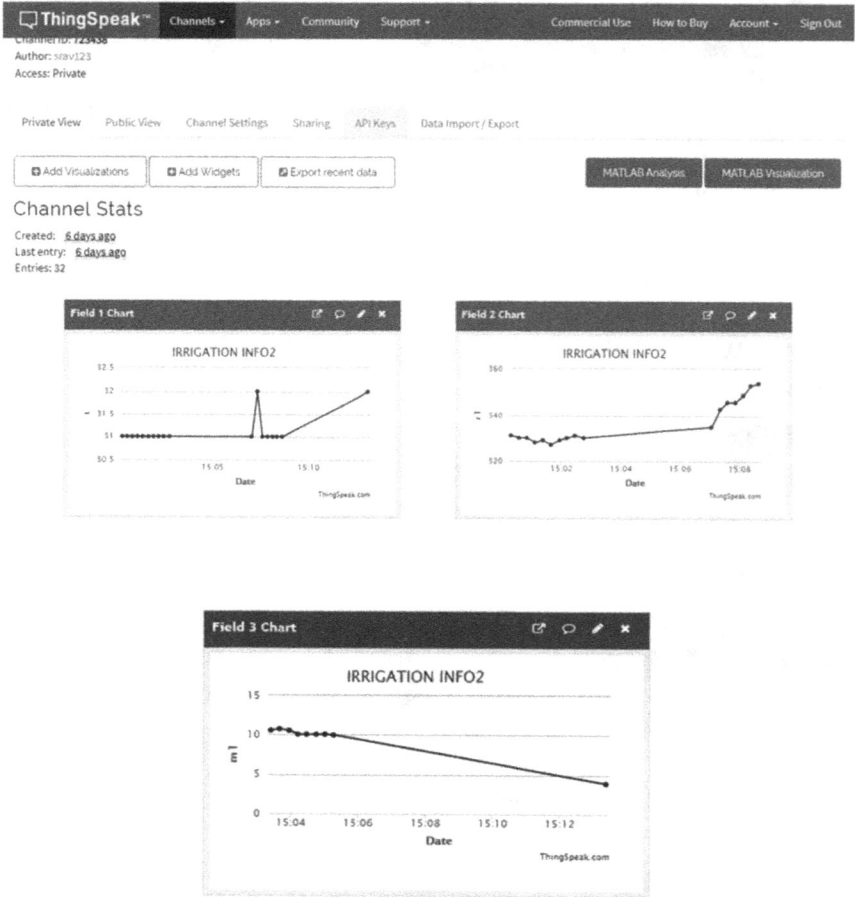

3.13   CONCLUSION

It is urgently imperative to have a program that supports the developing cycle and liberates the maker from troubles. With the ongoing innovative advances, the yearly generation of our country India, a totally agro-focal economy, should be expanded. One of the vital objectives of actualizing this advancement in the rural area is the chance to hold normal assets and to give a magnificent lift to the improvement of harvests. Vitality and time were the fundamental thought to spare the rancher's exertion. Frameworks should thusly be worked to successfully utilize the portable sensor systems, vitality sprinklers, GSM, SMS advancements to give this usefulness.

These frameworks were remotely controlled and given a minimal effort trade of data by means of SMS and GSM systems. Soil mugginess, humidity, and various other ecological elements impacting crop development are occasionally estimated utilizing an exact sensor of high caliber, and these qualities are transmitted to the processor to quantify the proper amount of water and composts and different explicit contributions during water system and provided in like manner to the rancher. GSM enables the programmed water system framework to expand efficiency by making it more comfortable to use through SMS, combined with missing activity.

The result of the study indicated the impact of GSM programming on farming water system frameworks and procedures being a correct methodology. The techniques examined have different points of interest and detriments in anticipation of strategies, trouble, and client commitment. New advances to additionally upgrade water system cycles, for example, prebuilt mobiles or incorporated water system process the executive's application programming, have been applied with the development progressing in regular day-to-day existence.

The "Checking SYSTEM AGRICULTURE MONITORING" can upgrade water levels dependent on soil humidity and climate expectations. This can be accomplished through remote humidity sensors that interface with the canny water system controls and enable the framework to tell whether the vegetation needs water or not.

Program Code:

```
#include <dht.h>
#include<stdlib.h>
#define baudrate 115200
#include <SoftwareSerial.h>
SoftwareSerial mySerial(4,5);
#include<LiquidCrystal.h>
LiquidCrystal lcd(2,3,9,10,11,13);
int contrast=50;
#include "DHT.h"
```

```
char phone_no[] = "9885352503";
char phone_no[] = "+91950";
bool temp = true;
bool wet = true;
bool dry = true;
bool sent = true;
bool motoron = false;
bool rain = true;
String cmd;
float c;
float permos1;
float permos2;
float permos3;
float rain1;
float rain2;
float rain3;
int t=0;
String GET1="GET https://api.thingspeak.com/
update?api_key=5G4BTA48NBS26N58&field1=";
String GET2="GET https://api.thingspeak.com/
update?api_key=5G4BTA48NBS26N58&field2=";
String GET3="GET https://api.thingspeak.com/
update?api_key=5G4BTA48NBS26N58&field3=";
#include <dht.h>
#include<stdlib.h>
#define ssid "ABCD"
#define pass"Haripriya@99"
#define api "api.thingspeak.com"
#define baudrate 115200
#define dhtpin 7
bool updated;
dht dht11;
void setup()
{
lcd.begin(16,2);
Serial.begin(baudrate);
analogWrite(12,contrast);
pinMode(A0, INPUT);
```

```
pinMode(A1, INPUT);
pinMode(A2, INPUT);
pinMode(A3, INPUT);
pinMode(A4, INPUT);
pinMode(A5, INPUT);
pinMode(8,OUTPUT);
mySerial.begin(9600);
}
void loop()
{
dht11.read11(dhtpin);
float c;
c=dht11.temperature;
delay(2000);
lcd.clear();
lcd.setCursor(0,0);
lcd.print("FINAL PROJECT");
if(isnan(c))
{
Serial.println("sensor not responding");
return;
}
Serial.print("temperature in celcius = ");
Serial.println(c);
float rain3=analogRead(A1);
float rain1 = analogRead(A3);
float rain2 = analogRead(A4);
float moist1 = analogRead(A5);
float moist2 = analogRead(A0);
float moist3 = analogRead(A2);
permos1 = (((1023 - moist1)/1023)*100);
permos2 = (((1023 - moist2)/1023)*100);
permos3 = (((1023 - moist3)/1023)*100);
Serial.print("moisture-1 = ");
Serial.println(permos1);
Serial.print("moisture-2 = ");
Serial.println(permos2);
Serial.print("moisture-3 = ");
```

```
Serial.println(permos3);
Serial.print("rain-1 = ");
Serial.println(rain1);
Serial.print("rain 2 = ");
Serial.println(rain2);
Serial.print("rain3= ");
Serial.println(rain3);
delay(2000);
if((permos1 < 31.5) && (permos2 < 31.5) && (permos3 < 31.5) &&
(rain2 > 700) && (rain1 > 700))
{
if((dry) || (!rain))
{
Serial.println("Ground is dry");
pinMode(7, HIGH);
delay(2000);
Serial.print("rain3= ");
Serial.println(rain3);
if(rain3<700)
{
lcd.clear();
lcd.setCursor(0,0);
lcd.print(" THE ACHEIVERS ");
lcd.setCursor(3,1);
lcd.print("MOTOR ON ");
t=0;
message1();
wifimsg();
motoron = true;
dry = false;
wet = true;
rain = true;
sent=true;
}
Else
{
pinMode(7, LOW);
lcd.clear();
```

```
lcd.setCursor(0,0);
lcd.print("FINAL PROJECT");
lcd.setCursor(0,1);
lcd.print("MOTOR NOTWORKING");
t=1;
message5();
wifimsg();
}}}
if(t==0)
{
if((rain1 < 400) || (rain2 < 400))
{
if(rain)
{
pinMode(7, LOW);
lcd.clear();
lcd.setCursor(0,0);
lcd.print("FINAL PROJECT");
lcd.setCursor(0,1);
lcd.print("RAIN:MOTOR OFF");
message4();
wifimsg();
rain = false;
dry = true;
}}
if((permos1 > 31.5) && (permos2 > 31.5) && (permos3 > 31.5))
{
if(wet)
{
Serial.println("Ground is wet");
pinMode(7, LOW);
lcd.clear();
lcd.setCursor(0,0);
lcd.print(" THE ACHEIVERS ");
lcd.setCursor(0,1);
lcd.print(" MOTOR OFF ");
message2();
wifimsg();
```

```
wet = false;
dry = true;
motoron = false;
sent=true;
}}
if(!motoron)
{
if(c > 26)
{
if(sent)
{
if(((permos1&&permos2)<31.5)||((permos3&&permos2) <31.5)||((permo
s1&&permos3)<31.5))
{
pinMode(7,HIGH);
lcd.setCursor(0,0);
lcd.print("FINAL PROJECT");
lcd.setCursor(0,1);
lcd.print("TEMP:MOTOR ON ");
message3();
wifimsg();
sent=false;
wet=true;
}}
}}
}}
void message1()
{
tone(8,1000);
delay(1000);
noTone(8);
delay(1000);
Serial.println("done");
delay(300);
Serial.println("AT+CMGF=1");
delay(2000);
Serial.print("AT+CMGS=\"");
Serial.print(phone_no);
```

```
Serial.write(0x22);
Serial.write(0x0D);
Serial.write
(0x0A);
delay(2000);
Serial.println("MOTOR TURNED ON");
Serial.print("temperature = ");
Serial.println(c);
Serial.print("first moisture percentage = ");
Serial.println(permos1);
Serial.print("Second moisture percentage = ");
Serial.println(permos2);
Serial.print("Third moisture percentage = ");
Serial.println(permos3);
delay(2000);
Serial.println (char(26));
}
void message2()
{
tone(8,1000);
delay(1000);
noTone(8);
delay(1000);
Serial.println("done");
delay(300);
Serial.println("AT+CMGF=1");
delay(2000);
Serial.print("AT+CMGS=\"");
Serial.print(phone_no);
Serial.write(0x22);
Serial.write(0x0D);
Serial.write(0x0A);
delay(2000);
Serial.println("MOTOR TURNED OFF");
Serial.print("temperature = ");
Serial.println(c);
Serial.print("first moisture percentage = ");
Serial.println(permos1);
```

```
Serial.print("Second moisture percentage = ");
Serial.println(permos2);
Serial.print("Third moisture percentage = ");
Serial.println(permos3);
delay(2000);
Serial.println (char(26));
}
void message3()
{
tone(8,1000);
delay(1000);
noTone(8);
delay(1000);
Serial.println("done");
delay(300);
Serial.println("AT+CMGF=1");
delay(2000);
Serial.print("AT+CMGS=\"");
Serial.print(phone_no);
Serial.write(0x22);
Serial.write(0x0D);
Serial.write(0x0A);
delay(2000);
Serial.println("TEMP:MOTOR ON");
Serial.print("temperature = ");
Serial.println(c);
Serial.print("first moisture percentage = ");
Serial.println(permos1);
Serial.print("Second moisture percentage = ");
Serial.println(permos2);
Serial.print("Third moisture percentage = ");
Serial.println(permos3);
delay(2000);
Serial.println (char(26));
tone(8,1000);
delay(1000);
noTone(8);
delay(1000);
```

```
}
void message4()
{
tone(8,1000);
delay(1000);
noTone(8);
delay(1000);
Serial.println("done");
delay(300);
Serial.println("AT+CMGF=1");
delay(2000);
Serial.print("AT+CMGS=\"");
Serial.print(phone_no);
Serial.write(0x22);
Serial.write(0x0D);
Serial.write(0x0A);
delay(2000);
Serial.print("HEAVY RAIN TURNING OFF THE MOTOR");
delay(1000);
Serial.println (char(26));
}
void message5()
{
tone(8,1000);
delay(1000);
noTone(8);
delay(1000);
Serial.println("done");
Serial.println("AT+CMGF=1");
delay(1000);
Serial.print("AT+CMGS=\"");
Serial.print(phone_no);
Serial.write(0x22);
Serial.write(0x0D);
Serial.write(0x0A);
delay(500);
Serial.print("FARM REQUIRES WATER");
Serial.print(" BUT MOTOR IS NOT WORKING");
```

```
delay(500);
Serial.println (char(26));
}
void wifimsg()
{
Serial.begin(baudrate);
Serial.println("AT");
delay(5000);
if(Serial.find("OK"))
{
bool connected=connectWiFi();
if(!connected)
{
Serial.println("WiFi is not connected");
}
Else
{
Serial.println("WiFiis responding");
}}
dht11.read11(dhtpin);
float t,h;
t=dht11.temperature;
Serial.println(t,h);
updated=updatevalues(String(t),String(rain1),String(permos1));
if(updated)
{
Serial.println("updated the values in cloud");
}
else
{
Serial.println("not updates");
}
delay(5000);
}
bool connectWiFi()
{
Serial.println("AT+CWMODE=1");
delay(2000);
```

```
String cmd="AT+CWJAP=\"";
cmd+=ssid;
cmd+="\",\"";
cmd+=pass;
cmd+="\"";
Serial.println(cmd);
delay(5000);
if(Serial.find("OK"))
{
return true;
}
else
{
return false;
}}
bool updatevalues(String temp,String r,String m)
{
String cmd="AT+CIPSTART=\"TCP\",\"";
cmd+=api;
cmd+="\",80";
Serial.println(cmd);
delay(2000);
if(Serial.find("Error"))
{
return false;
}
cmd=GET1;
cmd+=temp;
cmd+=GET2;
cmd+=r;
cmd=GET3;
cmd+=m;
cmd+="\r\n";
Serial.print("AT+CIPSEND=");
Serial.println(cmd.length());
if(Serial.find(">"))
{
Serial.print(cmd);
```

```
}
else
{
Serial.println("AT+CIPCLOSE");
}
if(Serial.find("OK"))
{
return true;
}
else
{
return false;
}}
```

KEYWORDS

- **Internet of things**
- **smart irrigation**
- **smart agriculture**
- **GSM**

REFERENCES

Blonquist, J. M.; Jones, S. B.; Robinson, D. A. Precise Irrigation Scheduling for Turfgrass Using a Subsurface Electromagnetic Soil Moisture Sensor. *Agric. Water Manage.* **2006,** *84,* 153–165.

Chavan, C. H.; Karande, P. Wireless Monitoring of Soil Moisture, Temperature and Humidity Using Zig Bee in Agriculture. *Int. J. Eng. Trends Technol. (IJETT)* **2014,** *11,* 1–5.

Dalip, V. Effect of Environmental Parameters on GSM and GPS. *Indian J. Sci. Technol.* **2014,** *7,* 1183–1188.

Davis, S. L.; Dukes, M. D. Irrigation Scheduling Performance by Evapotranspiration-Based Controllers. *Agric. Water Manage.* **2010,** *98,* 19–28.

Erdem, L. A. Y.; Erdem, T.; Polat, S.; Deveci, M.; Okursoy, H.; Gültas, H. T. Crop Water Stress Index for Assessing Irrigation Scheduling of Drip Irrigated Broccoli. *Agric. Water Manage.* **2010,** *98,* 148–156.

O'Shaughnessy, S. A.; Evett, S. R. Canopy Temperature Based System Effectively Schedules and Controls Center Pivot Irrigation of Cotton. *Agric. Water Manage.* **2010**, *97*, 1310–1316.

Grant, M. J. D. O. M.; Longbottom, H.; Atkinson, C. J. Irrigation Scheduling and Irrigation Systems: Optimising Irrigation Efficiency for Container Ornamental Shrubs. *Irrig. Sci.* **2009**, *27*, 139–152.

Idso, R. D. J. S. B.; Pinter, P. J. Jr.; Reginato, R. J.; Hatfield, J. L. Normalizing the Stress Degree Day Parameter for Environmental Variability. *Agric. Meteorol.* **1981**, *24*, 45–55.

Nemali, K. S.; van Iersel, Marc W. Anmicro System for Controlling Drought Stress and Irrigation in Potted Plants. *Sci. Horticul.* **2006**, *110*, 292–297.

Kim, Y. E. R.; Iversen, W. M. Remote Sensing and Control of an Irrigation System Using a Distributed Wireless Sensor Network. *IEEE Trans. Instr. Meas.* **2008**, *57*, 1379–1387.

Migliaccio, B. S. K.W.; Crane, J. H.; Davies, F. S. Plant Response to Evapo Transpiration and Soil Water Sensor Irrigation Scheduling Methods for Papaya Production in South Florida. *Agric. Water Manage.* **2010**, *97*, 1452–1460.

Nallani, S. B. H. V. Low Power Cost Effective Automatic Irrigation System. *Indian J. Sci. Technol.* **2015**, *8*, 1–6.

Prakash, M. G. U.; Ravichandran, T. A Smart Device Integrated with an Android for Alerting a Person's Health Condition; Internet of Things. *Indian J. Sci. Technol.* **2016**, *9*, 1–6.

Prateek Jain, P. K.; Palwalia, D. K. Irrigation Management System with Micro-Controller Application. *IEEE J.* **2017**. **DOI:** 10.1109/IEMENTECH.2017.8076969

Vijaya Kumar, S. V. D. Wireless Network Sensors for Precise Agriculture Monitoring. (IJIRSE). *Int. J. Innov. Res. Sci. Eng.* **2011**, *21*, 307–310.

Wang, W. Y. X.; Wheaton, A.; Cooley, N.; Moran, B. Efficient Registration of Optical and IR Images for Automatic Plant Water Stress Assessment. *Comput. Electron. Agric.* **2010**, *74*, 230–237.

Yuan, Y. L. G.; Sun, X.; Tang, D. Evaluation of a Crop Water Stress Index for Detecting Water Stress in Winter Wheat in the North China Plain. *Agric. Water Manage.* **2004**, *64*, 29–40.

CHAPTER 4

Attacks and Vulnerabilities Detection in Wireless Sensor Networks

RAKESH KUMAR SAINI[1,*], MOHIT KUMAR SAINI[2], and
RAVINDRA SHARMA[3]

[1]*School of Computing, DIT University, Dehradun, Uttarakhand, India*

[2]*Department of Computer Science, Doon Business School, Dehradun, Uttarakhand, India*

[3]*Swami Rama Himalayan University, Dehradun, Uttarakhand, India*

Corresponding author. E-mail: rakeshcool2008@gmail.com

ABSTRACT

The Internet of things (IoT) protects a big variety of productions and use occurrences that measure from an unattached restricted instrument as much as significant go-platform distributions of surrounded machinery and cloud constructions involving in real time. In this chapter, we describe operative service of IoT used for inspecting and controlling of standard home-produced situations with the support of little compensation instrument network. The involving instruments for responsible size of constraints via knowing instruments and broadcast of numbers via Internet are existence transported. This chapter presents a truncated cost and flexible home handle and watching device the practice of a surrounded hardware particle, with IP connectivity for having admission to and supervisory strategies and applications indistinctly the technique of mobile demonstration. The IoT is, at this moment, one of the maximum favorable equipment that have risen for many years. Wireless sensor networks are one of the leading supports for lots IoT communications. Wireless sensor networks and IoT devices are multiplied in lots of fields such as critical organizations including energy, transport, and industrial. Afterward, maximum of the unremarkable procedures now trust upon the information coming from sensors or IoT devices and their actions. The common vision of wireless sensor networks and IoT is usually associated through the uses of sensors. In IoT, sensors are used for detecting any movement in a reachable or not reachable area. This chapter strongly envisioned that effectiveness of existing Wi-Fi sensor networks (WSNs) can be improved by integrating IoT. In this chapter, we study the usage and development of wireless sensor networks inside the broader background of IoT and provide a working review of wireless sensor networks solicitations with IoT.

4.1 INTRODUCTION

A wireless sensor networks can normally be designated by way of a net of nodes that supportively intelligence and regulator the atmosphere, empowering communication among peoples or processers and the neighboring atmosphere. A wireless device community is a net shaped by way of a great variety of device bulges in which every bulge is prepared through an instrument to come crossways physical occurrences composed with light, warmness, strain, and so forth. Wireless sensor networks are appeared as a broad-minded information collecting technique to build the realities and discussion means so as to expressively improve the consistency and performance of organization constructions. Compared with the stressed solution, wireless sensor networks mouth informal positioning and better elasticity of strategies. Through the fast scientific growth of sensors, wireless sensor networks goes into the important thing generation for Internet of things (IoT). The net of issues is developing fast and in purpose human being's each day needs going to trust upon the net. It does not concern computers and smartphones anymore.[1,2] Multiple devices that we use daily life want the net to serve persons. The primary assistances of this broad sheet might be summarized as shadows: We observe Wi-Fi sensor networks (WSNs) and the Internet holistically, consistent with the imaginative and prescient in which WSNs may be a part of an IoT. Thereby, we pick out consultant utility eventualities for WSNs from the multidimensional WSN layout space, so one can gain insights into troubles involved with the mixing. In a wearable way like no other, the gesticulation-controlled armband senses muscle movement, so you can control any tool connected to the

IoT infrastructure just with your movement or gestures. The armband is ready with conductors to detect energy relaxation and hit upon discount and ruin of them when the hand is in undertaking. These movements are then forwarded to software at the backend that transforms and appreciates them into commands and achieves the task. Visualize being at the shoes of Tony Stark and justly pointing at your pill's screen to expose and nearby apps from far. The result is that no single structure will suit most of these regions and the requirements each region brings. However, a modular scalable structure that helps adding or subtracting abilities, as well as supporting many requirements across a huge form of these use instances is inherently useful and treasured. It affords a place to begin for architects trying to create IoT solutions in addition to a robust foundation for in addition improvement.[3] This chapter proposes this type of reference architecture. The reference architecture have to cowl a couple of elements including the cloud or server aspect architecture that allows us to display, manipulate, engage with, and process the information from the IoT gadgets; the networking model to talk with the gadgets; and the dealers and code at the devices themselves, in addition to the necessities on what form of device can support this reference architecture. The IoT mentions to the set of implements and constructions that communicate actual-worldwide instruments and actuators to the Internet. This includes many exceptional structures, together with internet-related cars and wearable devices that include fitness and health-tracking devices, watches, and even human implanted devices; smart meters and clever implements; home automation systems and lights controls smartphones that can be more and more getting used to amount the arena around them; and wireless sensor networks that degree weather, flood defenses, tides, and more. IoT also plays a vital role in business process improvements.[16] The growth of the wide diversity and style of devices, which might be gathering facts, is rather speedy. A look at by using Cisco approximations that the variety of internet-related implements passed the anthropoid populace in 2010, and that there might be 50 billion internet-related gadgets by 2020.[4,5]

4.2   WIRELESS SENSOR NETWORKS

In outdated wireless communication systems, the Open Systems Interconnection-layered building has been kind of surveyed and has attended many public services buildings nicely in the past, but developing wireless networks of these days are meaningfully tough this association attitude. The covered production designates a highland of technique layers wherein every layer cause inner its attractively distinct article and margin, and hereafter permitting rules to the essential expertise at each layer missing implementing the want to employment the overall system construction. This technique has been an achievement in its potential to provide modularity, clearness, and regulation personal the chain line structures but is in all probability misleading within the Wi-Fi structures location.[6,7]

Though wireless sensor networks, quit to end with cellular networks, Wi-Fi network function networks (WLANs), cellular statement-hoc networks (MANETs), and WSNs are comprehensively separate in positions of their plans and structure, a common topic in these varieties of networks is the use of the wireless channel for discussion. Different from the twine-line networks, the Wi-Fi channel has various unambiguous innovations that need to be taken seriously than reflected image at the same time as planning wireless networks.[8] Wireless sensor networks may have a position to play for some of navy operations besides enemy motion detection and stress tracking. Wireless sensor networks may be implemented for the aid of the merchant navy for a number of functions starting with modern awareness in far-off regions and energy protection. Being geared up with suitable sensors, these networks can stand discovery of player indication, evidence of identity of body weight, and investigation of their suggestion and development. The attentiveness of this text is at the competitive substances for plastic WSNs.[9,10] Based at the primary networking characteristics and navy use-cases, insight into particular army requirements is given on the manner to facilitate the reader's know-how of the operation of those networks in the near-to-medium term (in the subsequent 3–8 years).The chapter studies the evolution of military sensor networking devices with the encouragement of figuring out three generations of sensors along their abilities. Existing developer solutions are provided and a definition of a few current tailor-made produce for the navy environment is given.[11] The chapter accomplishes with an analysis of outstanding engineering and clinical difficult circumstances on the way to gain completely flexible, protection proved, advert hoc, self-organizing and mountable navy sensor networks. The term Conservational Instrument Nets has advanced to cowl numerous programs of wireless sensor networks to ground technology investigation. This contains identifying volcanoes, mountains, glaciers, forests, and so on. Some of these are indexed underneath. Area monitoring is commonplace software of WSNs. In region tracking, the WSN is deployed over an area in which a few singularities are to be supervised. An army example is the use of sensors to hit upon enemy intrusion; a civilian instance is the geo-fencing of gas or oil pipelines. A WSN includes spatially allotted independent sensors to show bodily or environmental situations, collectively with hotness, sound, vibration, stress, movement, or pollution and to helpfully skip their statistics via the net to a top region. The more cutting-edge networks are bi-directional, also permitting manage of sensor interest. The improvement of WSNs occurred by using navy packages in conjunction with battlefield surveillance; these days such networks are utilized in plenty of profitable and patron packages, together with industrial procedure tracking and management, gadget fitness monitoring, and so on.[12]

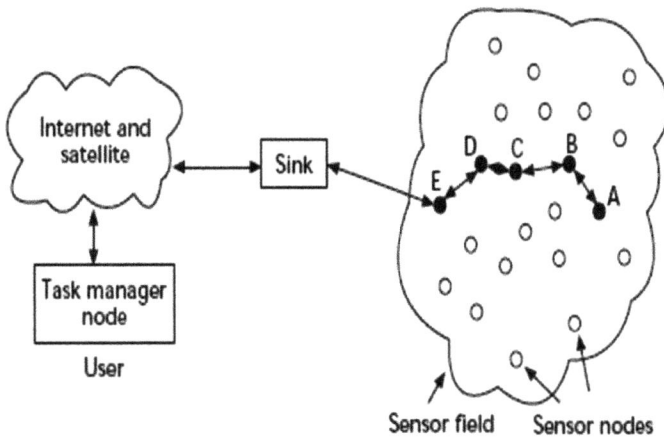

FIGURE 4.1   Architecture of wireless sensor network.

4.3   APPLICATIONS OF WSN

There are some popular applications of wireless sensor networks that are very useful for IoT:

1. Military
2. Environmental sensing

3. Area monitoring
4. Air pollution monitoring
5. Health
6. Forest fire detection
7. Home
8. Space exploration
9. Chemical processing

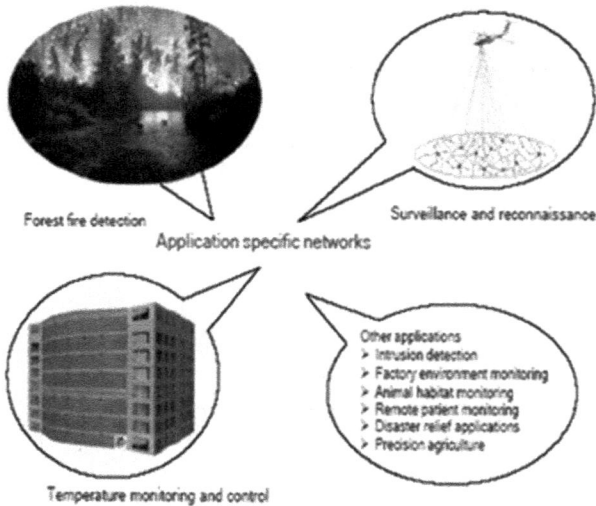

FIGURE 4.2   Application of wireless sensor network.

1. Military

Today wireless sensor networks have been used around the world. Smart sensors' important development in recent years has appeared. Response sensors are used for sensing, measurement, and data collection from atmosphere and transmission of the sensed data to each other or another user. The sensors are made as per their respective use. They will be economical than conservative sensors.[13,14]

2. Environmental Sensing

The period Conservation Device Systems has transformed to security numerous submissions of WSNs to powdered chastisement investigation. This contains recognizing volcanoes, mountains, glaciers, plantations, etc.

3. Area Monitoring

Area tracing is a conventional solicitation of WSNs. In district pursuing, the wireless sensor network is controlled over a constituency where some originality is to be managed. A military instance is consuming instruments to separate adversary obligation; a resident occurrence is the geo-fencing of petroleum or oil channels. Once the devices discover the occasion being observed (heat, strain), the occasion is mentioned to each of the improper positions, which formerly took appropriate motion (e.g., ship a message at the Internet or to a satellite). Similarly, WSNs can use various devices to perceive the occurrence of automobiles reaching from motorbikes to educate automobiles.

4. Air Pollution Monitoring

WSNs have been structured in numerous municipalities to demonstration the judgment of dangerous smokes for kingdoms.

5. Health

WSN technologies, with billions of dollars of funding and a worldwide network of manufacturers, are creating a health-care revolution. With a strong return on investment and excessive average revenue in keeping with consumer, there are dozens of health-care WSN "killer apps" for outpatient tracking, chronic ailment management, and geriatric care. While adoption of life-saving structures such as Wi-Fi ambulatory cardiac tracking has been progressively increasing over the last few years, the fastest developing WSN markets might be preventive health and well-being solutions. In 2012, wireless sensor solutions may want to store $25 billion international in annual health-care costs by way of reducing hospitalizations and increasing unbiased residing for seniors.[15]

6. Forest Fire Detection

WSNs may be established in forests to control wildfires. The nodes might be geared up with instruments to operate high temperature, moisture, and smokes that can be produced by means of fire inside the trees or vegetation. The primary discovery is critical for a success undertaking of the firefighters; appreciations to WSNs, the fireplace group resolve capable of realize while a fireplace is began and how it's miles spreading.[26]

7. Home

Standards-based completely Wi-Fi sensor municipal implementation confidential the home has extensive been different with home mechanization systems. Progressively wireless sensor networks will arranged inside those constructions as the knowledge fits the duplicate extensive national application expanses that home mechanization has long maintained–energy switch, home entertaining operate, health care, and security, but they will more overhead a better ad-hoc, patron-led deployment, and adoption of wireless automation inside homes. Driving that is developing implementation of desired WSNs era. Regularization conveys extremely increase the size of the addressable market in addition to reducing fees for allowing wireless communications beginning up the potential for large interoperability across multiple gadgets. Home wireless sensor networks uptake will pressure exchange inside home automation as well as across more than domestic control, amusement, and purchaser electronics gadgets. The broad potential for bringing Wi-Fi control to quite a number home devices will draw players from several businesses to proposal connectivity in addition to guide for the allowed correspondences.[17]

8. Space Exploration

We can inspect the examination of the stellar device by resources of ad-hoc WSNs, that is, networks, wherein all nodes propose and convey information. The features of self-business innovativeness and localization are the leading experiments to overcome to obtain a consistent network for a feast of missions. We aspect out the variety of conservational and working forces that might must expression wireless sensor networks used for space exploration. The first organization of eventualities we evaluated studies nodes transferring relative to each other either above or on the floor of a sun system object. These possibilities permit collecting records concurrently over an enormous ground. The second group of situations we occupied into thought studies the use of nodes constant in or on the ground of an asteroid or planet. We considered both physical and chemical sensing of the atmosphere, floor ground, and soil as applicants for such networks. Emerging extremely advanced technologies are examined so that you can make a distinction between the elements that can be commonplace for a spread of missions and the others which might be precise to an examination state of businesses.

Finally, we inspect the unique requirements of wireless sensor networks for universe exploration with those of wireless sensor networks calculated for earthly correspondences.[18,19]

9. Chemical Processing

Directing at pursuing histories of HCHO, CH_4, LPG and dissimilar enormous and damaging petroleum attentions in the substance industry, the following machine designed a Zigbee WSNs self-possessed of CC2430, MC114, and MS1100 gasoline sensor, which constructed a secure, low-electricity-intake, bendy detection device for harmless fuel checking. As the device information broadcast with the method of sensor takes a look at information in a rapid remoteness, multi-node, personality-forming wireless network, it can use the USB interface gateway to attach the wireless sensor networks and Internet to apply the evidences broadcast with the Internet.

4.4   INTERNET OF THINGS (IOT) VISION

The dream of the IoT may be unstated from insights—"Internet-centric" and "detail-centric." The Internet-centric construction includes internet conveniences because the principle emphasis, as records is being shaped by using the "matters." In the thing-centric structure, smart gadgets take the center degree. Equinox is at the coronary heart of the interconnected worldwide of the Internet, so we are targeted on the internet-centric view. Today we live in an international wherein globally the call of the game for all a fulfillment organizations is innovation; revolution within the framework of computing and communiqué is more and more most important to a convergence of the arena of computer systems with the natural global. Artificial Neural Network Detect and recognize activities of daily life.[25] This manner is a growing quantity of major to the IoT or the civilization of effects. Envisioning global of IoT in 2020 allows companies recognize rising IoT possibilities and plan for functionality building.[21] The 1-day event titled Vision 2020 with The IoT makes a place of expertise of three critical apprehension subject matters—The Manufacturing Smart Industry, The Clever Urban, and The Clever State.

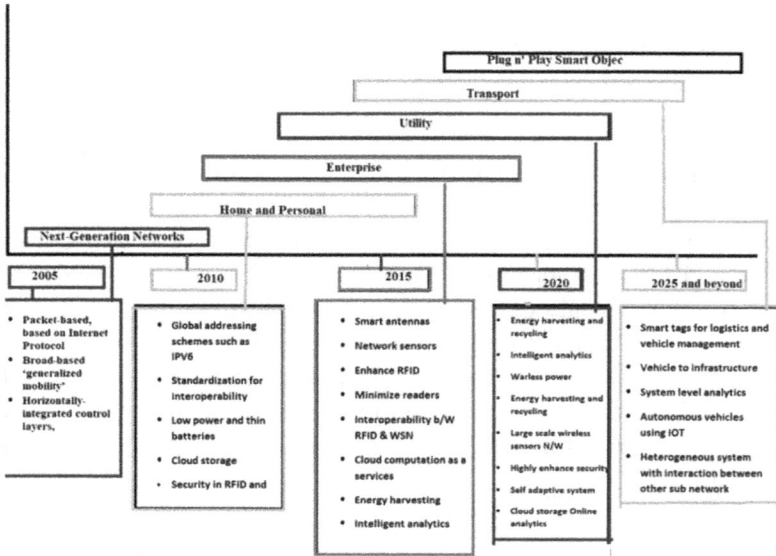

FIGURE 4.3   Internet of things (IoT): A vision, architectural elements, and future directions. *Source*: Self made image (https://www.sciencedirect.com/science/article/abs/pii/S0167739X13000241) taken from figure 8, roadmap of key technological developments in the context of IoT application domains envisioned)

4.5   INTERNET OF THINGS ARCHITECTURE

The house of IoT comprises a fixed of mechanisms. Layers can be set up out via definite technology, and we are able to talk choices for understanding every matter. There are also a few pass-slicing/upright layers that comprise get-admission-to identification control. The efficiency and applicability of any such machine directly associate with the first-rate of its building blocks and the way they have communication, and there are numerous methods to IoT building. In this chapter, our IoT accessing professionals will share their arms-on enjoy and gift their innovative concept of a mountable and bendy IoT structure.[20,21]

4.5.1   CODING LAYER

Coding coating is the basis of IoT which delivers certification to the things of awareness. In this layer, respectively, article is allocated an exclusive ID which types it informal to differentiate the constituent.

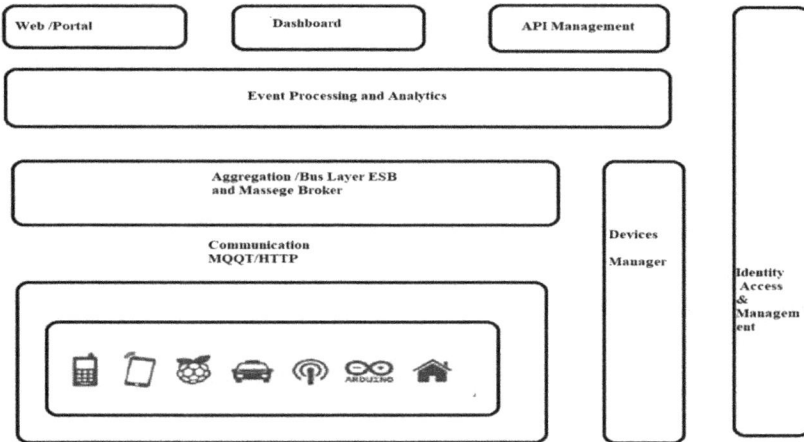

FIGURE 4.4   Architecture of IoT.

4.5.2   PERCEPTION LAYER

It is the key simple share of IoT. The investigation of device lump is the current spot each home and remote. And that may supply remarkable loss. This is the tool layer of IoT that offers a corporeal significance to each object. It consists of data sensors in unalike bureaucracy like RFID tags, IR sensors before dissimilar sensor networks that could intelligence the temperature, stickiness, speed and region, and so on. This layer gathers the useful facts of the items from the sensor devices associated with them and translates the information into normal in dictations that's then handed onto the Network Layer for added achievement.[21,23]

4.5.3   NETWORK LAYER

It is the unstable part inside the entire IoT system. There can be one-of-a-kind production community connected to each different. The honesty man or woman of IoT makes it face many individuality evidence safety issues. In adding, single of the frequent appearances of IoT is the enormous quantity of information. When sensor nodes understand, they produce an enormous number of needless figures necessarily, in an effort to reasons network crowding inside the technique of broadcast. And this is in all likelihood to generate denial of provider assaults. So we need to upload the filtration devices among the transmission layer and the utility layer to ensure the community cleared. Performs the following capabilities; Gateway—Routing and Addressing—Network Capabilities—Transport Capabilities—Error detection and Correction.[24]

4.5.4 MIDDLEWARE LAYER

These coating procedures the statistics installed after the device strategies. It consists of the technology like Cloud figuring, Universal calculating which safeguards an instantaneous get admission to the database to shop all the vital data in it. Using a few intellectual dispensation gears, the data is processed and a totally computerized movement is in use primarily founded on the treated outcomes of the statistics.

4.5.5 APPLICATION LAYER

As per is acknowledged to us, software is the reason of emerging IoT. It creates our lifetime increase extra intellectual and decreases our undertaking. Though, within the solicitation procedure of an exact manufacturing, perceptive layer gathers big quantities of facts of workers, counting approximately confidentiality data.[9] Consequently, the way to shield those significant confidentiality facts connecting man or woman, companies or situations is the huge sum of software layers, including the IoT utility. This layer is at the highest of the construction and is actionable for shipping of various packages to one-of-a-kind customers in IoT. The applications can be from specific industry segments such as environment, public protection, production, logistics, retail, health care, meals, and drug. With the growing maturity of RFID era, numerous packages are evolving, so one can be below the authority of IoT. This layer understands the programs of IoT for all kind of enterprise, based at the treated facts. Because packages sell the development of IoT, this layer may be very useful inside the big degree growth of IoT community. The IoT-associated applications could be clever transportation, smart houses, clever planet, and many others.

4.5.6   BUSINESS LAYER

The business shelter accomplishes the entire IoT system, including applications, business and income models, and users' privacy. This coating completes the requests and facilities of IoT and is answerable for all the investigation related to IoT. It produces dissimilar occupation a representations for operative commercial approaches.

4.6   WIRELESS SENSOR NETWORKS VS INTERNET OF THINGS

WSNs refer to a group of particular dedicated sensor with a communications infrastructure. WSNs are the base of IoT applications. It is used for observing and recording the corporal situation. IoT is the network of physical objects, and it is used for connect and replace data over the Internet. In simple words, IoT is concept of things interaction with the Internet. IoT systems direct their information to the Internet. IoT uses direction-finding to direct their data to the Internet. In 2019, there may be regularly confusion in articles regarding the distinction between IoT and wireless sensor networks, often inflicting them to be vague collectively. In this text, we are able to go through some of the foremost difference between these two standards and clarify how they may be, in fact, pretty exclusively alike. Knowledge developments in Wi-Fi infrastructures and microelectronics have allowed the growth of low-cost, low-strength, multipurpose sensor nodes that might be tiny in length and talk in brief distances. These small and normally simple sensor nodes include sensing units, data processing, and speaking additives. A large number of such nodes deployed over big areas can paintings collectively with each different. To be gainful, the sensor nodes are often kept in very confined electricity reserves. Early strength discount can critically restrict the network carrier and wishes to be addressed considering the IoT software need for cost, deployment, renovation, and carrier availability.

4.6.1   INTERNET CONNECTIVITY

In an IoT machine, all the sensors directly throw their information to the net. For instance, a sensor may be used to take a look at the temperature of a body of water. In this situation, the statistics could be immediately or periodically dispatched immediately to the net, in which a server can procedure the records and it is able to be interpret on a front-end interface. Conversely, in a WSN, there is no direct connection to the net. Instead, the diverse sensors attach to some sort of router or imperative node. A character can also then path the facts from the router or critical node as they see fit. That being said, an IoT device can utilize a wireless sensor network via communicating with its router to collect facts.

4.6.2   WSN AS A SUBSET OF IOT

IoT exists at a better level than WSN. In other phrases, WSN is usually an era used within an IoT system. A huge collection of sensors, as in a mesh community, may be used to in my view gather records and throw information thru a router to the Internet in an IoT machine. It's crucial to note that the term "wireless sensor community" is not always nearly as round as "the Internet of things." WSN consists of a community of most effective wireless sensors. If the network became to encompass a stressed out sensor, it may no longer be categorized a "wireless sensor community." This is awesome IoT. Basically any device that connects to the net may be taken into consideration an IoT tool. An "Internet of Things system" can consequently be interpreted as a collection of many IoT devices.

Examples

A fridge with the high-temperature know-how to the Internet is amazing to use as a wireless sensor network but it is an IoT tool. A series of sensors used to monitor precipitation on an acre of land is possible to be taken into consideration a "wireless sensor network" if in truth all the sensors are wireless.

4.7   APPLICATION REQUIREMENTS OF WI-FI SENSOR NETWORK FOR INTERNET OF THINGS

The recognizing lumps talk in multi-hop every sensor is a transceiver has an antenna, a microcontroller, and an interfacing circuit for the sensors as a communication, actuation and sensing unit, respectively, in conjunction with a source of electricity that can be both battery and any electricity harvesting era. By using this project a supplementary part for exchangeable the facts, named as Memory Unit which can additionally be a portion of the detecting lump. A regular sensing node is exposed inside it:

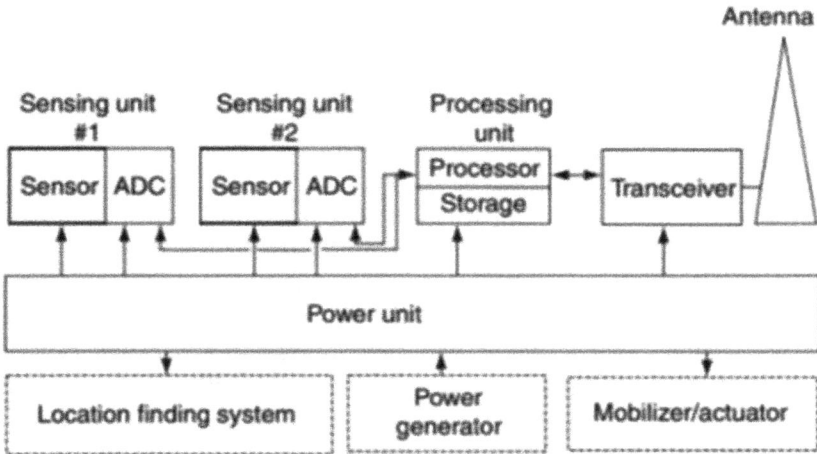

FIGURE 4.5 Architecture of wireless sensor networks.

Figure 4.4 has proven the gears of sensor node. There are four components of sensor nodes are power unit, processing unit, storage unit, and transceiver unit. In the architecture of wireless sensor networks, sensors sense the environment from its surrounding location and convert this records analog to virtual and transfer to processing unit from in which statistics sensor switch to transceiver unit. Wireless Sensors Network era and RFID generation while joint together opens up opportunities for even extra stylish gadgets, for which a number of solutions have been projected. Moreover, most of the WSNs are primarily based at the IEEE 802.15.4 standard, which specify the Physical and MAC layer of Low-Rate Wireless Personal Area Networks (LR-WPANs). The technologies that enable the mixture of WSN with the IoT are a hot research subject matter, many answers had been projected for that which include that of a 6LOWPAN well-known, that permits IPv6 packets to be transmitted in the course of the networks which might be computationally restrained. Also there's ROLL routing fashionable for quit-to-give up routing solutions. Our decision is to lower energy eat-through sensor and enhance its performance. Sensor accumulate the massive quantity of facts from scenario which we are storage in database person constantly seeking out short and essential records or records from database so our goal is to fulfill user's forecast by means of using information mining algorithm for getting access to statistics from database. Data mining is method used to mine quick and vital facts from big amount of records. Dependability the sensor continually in lively nation compulsory large amount of energy so as to decrease this power gobbling we are able to transfer sensor from energetic to shiftless and shiftless to energetic kingdom as consistent with person's request. Systems have to take selections beyond its limits. That is, the system must behave rationally. Sensors each day collect the statistics and store it in the cloud storage. The cloud storage is to be accessed only to the authorized person. Someone should not modify or sell the records within the cloud. We are going to use strong affirmation technique for this cause. Most of the day-by-day life requests that we usually see are already clever; however, they are unable to interconnect with each other and permitting them to interconnect with each different and proportion valuable information with each other will create a huge style of unique packages. There may be a proper tracking of air pollutants in the surroundings. Hospitals will be ready with smart supply wearable embedded with RFID tags if you want to receive to the patients on arrivals, thru which now not simply medical doctors but nurses may also be capable to screen heart rate, blood strain, temperature, and different situations of sufferers internal or outside the constructing of clinic.

4.8 CONCLUSION

Wireless sensor networks or even greater so, IoT, are not separated expertise but instead characterize composite systems the procedure of several knowledge from bodily communiqué layers to solicitation programmers and stand charity in numerous charge expanses and distinctive atmospheres. In this chapter, we investigated the IoT–enabled technology in terms of shrewd metropolises, heterogeneous IoT, fog computing, data mining, WSN-based facts centric IoT, cellular communiqué, context-attention, virtualization, and real-time analytics. To comprehend the inspirations of utilizing diverse IoT components, we brought the necessities of different IoT factors with their time-honored goals. Next, we supplied an examination of classes found out from different research that have been reviewed at some point of this chapter. To conclude, approximately exposed research-demanding situations associated with the stated regions were furthermore discussed for the instinct of the IoT acceptability. In this chapter, the presented architecture of wireless sensor networks and IoT are simple and can be easily accepted for similar deployments. In this chapter, it is clear that the potential of the wireless sensor networks paradigm will be fully unbridled once it is connected to the Internet, becoming part of the IoT.

KEYWORDS

- **Internet of things**
- **wireless sensor networks**
- **monitoring platform**
- **home automation**
- **wireless sensor nodes**

REFERENCES

1. Akyildiz, I. F.; Su, W.; Sankarasubramaniam, Y.; Cayirci, E. A Survey on Sensor Networks. *IEEE Commun. Mag.* **2002,** 102–114 (August).
2. Ferrari, P.; Flammini, A.; Marioli, D.; Sisinni, E.; Taroni, A. Wired and Wi-Fi Sensor Networks for Business Packages. *Microelectr. J.* **2009,** *40* (9), 1322–1336.
3. Ferrari, P.; Flammini, A.; Rizzi, M.; Sisinni, E. Improving Simulation of Wi-Fi Networked Manage Systems Based on Wireless HART. *Comput. Stand. Interf.* **2013,** *35* (6), 605–615.
4. Dash, A. K.; Mohapatra, S.; Pattnaik, P. K. A Survey on Application of Wireless Sensor Community the Usage of Cloud Computing. *IJCSET* **2010,** *1* (4), 50–55.
5. Durisic, M. P.; Tafa, Z.; Dimic, G.; Milutinovic, V. In *A Survey of Military Packages of Wireless Sensor Networks*, 2012 Mediterranean Conference on Embedded Computing (MECO); 19–21 Jun 2012; pp 196, 199.
6. Depari, A.; Flammini, A.; Sisinni, E.; Vezzoli, A. In *A Wearable Smartphone-Based System for Electrocardiogram Acquisition*, 2014 IEEE International Symposium on Medical Measurements and Applications Proceedings; Lisbon, Portugal, 11–12 Jun 2014; pp 54–59.
7. Kaghyan, S.; Sarukhanyan, H. Accelerometer and GPS Sensor Mixture Primarily Based Gadget for Human Activity Recognition. *Comput. Sci. Info. Technol. (CSIT)* **2013,** 1–9.
8. Zhang, Z.; Tiejun L. v; Su, X.; Gao, H. In *Dual Xor Inside the Air: A Community Coding Based Totally Retransmission Scheme for Wireless Broadcast Broadcasting*,

Communications (ICC), 2011 IEEE International Conference on; IEEE, 2011; pp 1–6.

9. Al-Fuqaha, A.; Guizani, M.; Mohammadi, M.; Aledhari, M.; Ayyash, M. Internet of Things: A Survey on Allowing Technologies, Protocols, and Packages. *Commun. Surv. Tutor., IEEE* **2015**, *17* (4), 2347–2376.

10. Shen, C.; Srisathapornphat, C.; Jaikaeo, C. Sensor Information Networking Architecture and Applications. *IEEE Pers. Commun.*, Aug **2001**, 52–59.

11. Shih, E. et al. In *Physical Layer Driven Protocol and Algorithm Design for Energy-Efficient Wireless Sensor Networks*, Proc. ACM MobiCom '01; Rome, Italy, July 2001; pp 272–286.

12. Wang, H.; Yip, L.; Maniezzo, D.; Chen, J.; Hudson, R.; Elson, J.; Yao, K. In *A Wi-Fi Time Synchronized Cots Sensor Platform Component II–packages to Beamforming*, Proceedings of IEEE CAS Workshop on Wireless Communications and Networking; Pasadena, CA, 2002.

13. Wang, H.; Elson, J.; Girod, L.; Estrin, D.; Yao, K. In *Target Class and Localization in Habitat Monitoring*, Proceedings of the IEEE ICASSP 2003; Hong Kong, April 2003.

14. Wang, H.; Estrin, D.; Girod, L. Preprocessing in a Tiered Sensor Community for Habitat Monitoring.

15. Xu, Y.; Bien, S.; Mori, Y.; Heidemann, J.; Estrin, D. *Topology Control Protocols to Conserve Energy in Wi-Fi Advert Hoc Networks*; Technical Report 6, University of California, Los Angeles, Center for Embedded Networked Computing, January 2003. Submitted for Booklet.

16. Rana, G.; Sharma, R. Emerging Human Resource Management Practices in Industry 4.0. *Strat. HR Rev.* **2019**, *18* (4), 176–181.

17. Alkar, A. Z.; Buhur, U. An Internet Primarily Based Wi-Fi Home Automation Device for Multifunctional Devices. *IEEE Trans. on Consumer Electr.* **2005**, *51*, 1169–1174.

18. Sharma, U.; Reddy, S. R. N. Design of Home/Office Automation Using Wireless Sensor Network. *Int. J. Comput. Appl.* **2012**, *43*, 53–60.

19. Liang, N.-S.; Fu, L.-C.; Wu, C.-L. In *An Incorporated, Flexible, and Internet-based Control Structure for Domestic Automation Machine in the Internet Generation*, Robotics and Automation, 2002. Proceedings.ICRA'02.IEEE International Conference; 2002; pp 1101–1106.

20. Rajabzadeh, A.; Manashty, A. R.; Jahromi, Z. F. A Mobile Application for Smart House Remote Control System. *World Acad. Sci., Eng. Technol.* **2010**, *62*, 80–86.

21. https://shiverware.com/iot/iot-vs-wsn.html.

22. Ashton, K. That 'Internet of Things' Thing. In the Real World, Things Matter More Than Ideas. *RFID J.*, 22 June 2009. http://www.rfi djournal.com/articles/view?4986.

23. Hatler, M.; Gurganious, D.; Chi, C. *Industrial Wireless Sensor Networks. A Market Dynamics Report*; ON World, 2012.

24. Sen, J. A Survey on Wireless Sensor Network Security. *Int. J. Commun. Netw. Info. Security (IJCNIS)* **2009**, *1* (2), 5578. http://arxiv.org/ftp/arxiv/papers/1011/1011.1529.pdf.

25. Singh, R.; Anita, G.; Capoor, S.; Rana, G.; Sharma, R.; Agarwal, S. Internet of Things Enabled Robot Based Smart Room Automation and Localization System. In *Internet of Things and Big Data Analytics for Smart Generation*; Balas, V., Solanki, V.,

Kumar, R., Khari, M., Eds.; Intelligent Systems Reference Library; Springer: Cham, **2019,** *154.*

26. Heidemann, J.; Xu, Y.; Estrin, D. In *Geography-Knowledgeable Energy Conservation for Advert Hoc Routing,* Proceedings of the Seventh Annual ACM/IEEE International Conference on Mobile Computing and Networking (Mobicom 2001); Rome, Italy, July 2001.

CHAPTER 5

Agricultural Science with IoT

YOGESH PANT*

Department of Computer and Information Studies, Himalayan Institute of Science and Technology, SRHU, Dehradun, Uttarakhand, India

*E-mail: yogepant123@gmail.com

ABSTRACT

Today, the world is rising in the field of technology and the basic need of human being is food, home, and cloths, which are directly or indirectly fulfilled by agriculture. Incorporation of technology with agriculture will increase the production and automation of process and reduce human intervention and efforts. There are lots of technologies that are helping us to improve the methods of agriculture such as intelligent system for weather forecasting and microcontroller-based equipment for automation of process and many more. These technologies have their advantages and also improving the production.

Internet of things (IoT) is an emerging technology that incorporates intelligent system, cloud computing, and microcontroller itself. In this chapter, we will go through a deep drive about the things (sensors and actuators) that are connected with cloud system and the cloud computing platforms that are suitable to use for IoT projects with their features. In agricultural revolution, information technologies such as IoT, cloud computing, big data, and machine learning are playing a vital role. Different sensor modules such as soil sensor, moisture sensor, humidity sensor, temperature sensor, and many other sensors are useful for agriculture science that measures the environment variables and water pump, pesticide sprinkler, cold/hot air blower, and temperature control units are the actuators for different types of indoor and outdoor crops.

Microcontroller selection is the main challenge for implementation of IoT-based system. In this chapter, we will also cover different microcontroller and there uses as per application scope, power consumption, storage, and network connectivity and will implement IoT module on basil plant and collect the threshold data. There are different levels of IoT applications and development methodology and IoT project design reference model. Depending on the levels of IoT application, we decide the location for storage unit and computational unit, and the design reference model is used make plan and prepare resources before developing and deploying an IoT system. Sensors and actuators are connected with microcontroller, sensor collects the environment data and sends to microcontroller that uses the network routers to transmit the data to the cloud through gateway nodes and then computational unit is responsible for triggering the action using actuator devices. Communication between microcontroller to storage and computational unit can be wired or wireless. There are a number of cloud platforms developed for IoT-based systems. Using cloud we connect the end user via mobile application, web application, or hardware controller. These applications provide user interface for monitoring the sensor data and also can trigger the actuators to perform action.

5.1   INTRODUCTION

Agriculture is one of the oldest fields of human involvement; we are always freaking around the agricultural products in mostly every form either in the form of food, clothes, medicine, or any other day-to-day needs. We do not exactly know the time when human have started growing food because the oldest literature that we have from that time we have traces of agriculture. We have different types of environment around the world, which is directly responsible for diversity in agriculture around the world. There are environmental conditions and soil capability that nurture different types of crops in different places and in different months of the years or in different seasons.

We can manage to grow any crop at any place or time; only we need to maintain the environmental conditions and the nutrition in the soil. Agriculture science studies about needs of environmental parameters such as temperature, humidity, pressure, and the most important thing is the medium for growing the crop, soil.

Nowadays we can grow any crop in land in traditional or hydroponic methods in any season only by controlling the environmental factors that are connected with the particular crop, and the technology is going to provide support for this. Today we not only can grow the different crops in any time but also we can reduce or remove human intervention for cultivation. The automation systems which are based on electronic system are the pulse of modern technology. They sense, compute, and control real-world environment based on user needs.

The emerging technologies such as Internet of things (IoT) and cloud-based platform are going to change the way of agriculture that we are using traditionally, which will improve the production, and at the same time, it will reduce the requirement of manpower. IoT is converting the agriculture into smart agriculture industry. This works in different domains of agriculture such as water management, temperature and humidity control, soil management, crop monitoring, and control of insecticide and pesticides. All these factors are going to increase the production and profit. The main benefit of technology such as IoT in agriculture is that we can visualize the parameters and control them from anywhere in the world. That provides a centralized control and visualization interface on the user's hand for geographically different agricultural centers. Cloud computing is an emerging field of information technology, that is the base platform for trending technologies such as machine learning, IoT, big data, and social networking. Cloud computing is the best business opportunity for the investors because the services of cloud computing is everywhere such as online marketing, end-to-end customer connectivity and automatic monitoring of supply chain for farmer's also. IoT is the technology that works over the network with good bandwidth; this is why data storage and data analytics is being performed over cloud system. There are so many different types of services for the farmers which IoT provides such as smart greenhouse system, agricultural equipment that are enabled with IoT, and drones that are used for crop monitoring and control of insecticide and pesticides.

5.2   INTERNET OF THINGS

IoT is dynamic global internetwork with self-configuring capabilities of devices that could be electrical or mechanical, and they can be identified uniquely. For example, LED light, vending machine, water sprinkler for plants, body sensors, and driverless cars are application of IoT. IoT is based on electronic system enabled with the Internet. The real-world parameters are analog, while the computations performed within the processing system are in digital form. Hence the electronic system involves both digital and analog interfaces for the signal flow between real-world interface and processing system. There are different parts of IoT system. First is the embedded system that covers hardware part and involves sensors, micro-controllers, and connectivity modules, and the second is the software part that is the programming language that will program the microcontroller and another is the Internet for providing connectivity between sensors, actuators, and the base station.

Typical signal flow chain starts from the real world and the parameters such as temperature, position, humidity, speed, flow, light, and sound are received into amplifier and then the analog data converts into digital using data converter and then the data passed through embedded processing that implements logic and then appropriate signal send to actuators.

The scope of IoT is not only to connect the things, that is, devices or machines, to the Internet. IoT provides a communication channel by which they can communicate and exchange the data. A typical IoT system has things or nodes connected with a web server or cloud and people who can access and monitor the data over the Internet.

5.3   IOT APPLICATION LEVELS

IoT applications have different levels based on the complexity of elements and complexity of the application. In IoT application, there are nodes sensors or actuators, routers or gateway for connectivity, and a data analysis center that analyze the data and trigger the nodes as per logic. Following are the different levels of IoT applications and their major features.

TABLE 5.1   IoT Project Levels and Respective Features.

Levels	Features
1	• Single node connected with cloud
	• Local data storage and performance analysis
	• No complex analysis

TABLE 5.1   *(Continued)*

Levels	Features
2	• Single node
	• Data storage on cloud
	• Monitoring node connected with cloud
	• Big data and no complex analysis
3	• Single node
	• Data storage and analysis on cloud
	• Monitoring node connected with cloud
	• Big data and complex analysis
4	• Multiple nodes
	• Data storage and analysis on cloud
	• Monitoring node connected with cloud
	• Big data and complex analysis
5	• Multiple nodes with local coordinator
	• End points connected with local coordinator
	• Local coordinator communicate with cloud via the Internet
	• Big data and complex analysis
	• Observer node on cloud makes it intelligent IoT
6	• Multiple nodes with remote coordinator
	• Centralized controller between monitoring nodes and cloud
	• Data storage and analysis on cloud
	• Big data and complex analysis
	• Observer node are local as well as cloud

5.4   IOT SYSTEM DESIGN REFERENCE MODEL

System with a development strategy results successful and goal-oriented system. When we design an IoT system, we have to follow a certain strategy that is known as IoT system design reference model. System design reference model comes with a seven-step process:

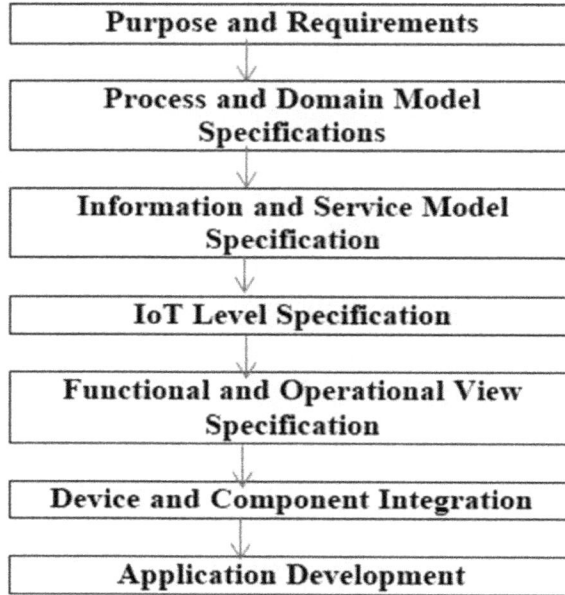

FIGURE 5.1   IoT system design reference model.

We will see an example and understand the activity involved in each step.

Example—Automated Garden Sprinkler

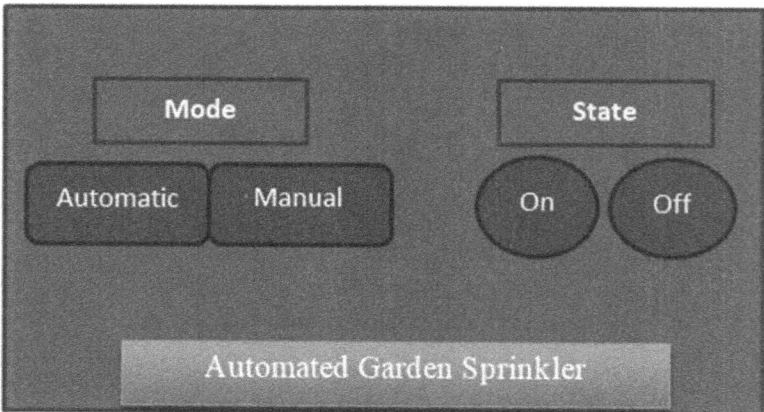

FIGURE 5.2   User interface for automated garden sprinkler.

Above is the user interface from which user can select the mode of sprinkler either automatic or manual and also can set or check the state of the sprinkler on and off. Now we go through system design and reference model for automated garden sprinkler.

1. Purpose and requirements

    i.   Purpose—Automated garden sprinkler that allows controlling sprinkler remotely using a web application.

    ii.   Behavior—Automated garden sprinkler has automatic and manual mode. In automatic mode, microcontroller system will measure the humidity in the soil and turn on the water sprinkler when humidity is below the threshold value. In manual mode, we can turn on and off the water sprinkler remotely.

    iii.   Application development requirements—Application for collecting, analyzing the data and application that will run on user end that provides dashboard to get the sensor data and provides remote access control over the devices.

    iv.   Implementation requirement—Implementation requirement deals with the hardware modules such as microcontroller and network devices as well as the cloud platform.

    v.   Security requirements—The system should be capable to authenticate the user using username and password. Security over the network is very important and research-oriented field in the IoT.

    vi.   System management functions—The system should provide user interface for remotely monitoring the status of humidity level and buttons to trigger the sprinkler.

    vii. Data analysis requirement—We have data in the form of humidity level that can microcontroller manage locally.

2. Process model specification
3. Domain model specification

    i.   Entity—Humidity level is an entity

    ii.   Resources—Electricity and water are the resources

    iii.   Devices—Microcontroller, sensors, and water pump are the devices

    iv.   Services—Remote access and controlling the entity using the devices

4. Information model specification
   i. List down different entities—we have only one entity—Humidity
   ii. Associate attribute and relationship with entities (if any)
5. Service specification
   i. List down different required services
      a. Remote access of status—Humidity level
      b. Remote controlling the devices
6. IoT level specification
   Based on domain and service model specification, this application comes under Level 1.
7. Functional view specification
   Prepare the functional view diagram for better understanding of the system.
8. Operational view specification
   i. Devices—Select microcontroller
   ii. Communication protocols as per TCP/IP model
   iii. Services—Remote monitoring and controlling
   iv. Application—Web application or mobile application
   v. Security—Authentication
9. Device and component integration—Build an IoT system
10. Application development—Integrate IoT system with use interface

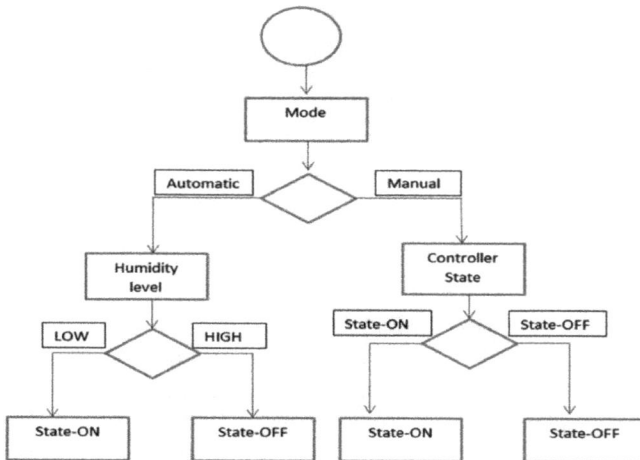

FIGURE 5.3   Process model specification of automated garden sprinkler.

5.5   IOT DEVICES

There are many companies that are providing IoT hardware and also integrated cloud service for IoT solutions. It is very important task to select appropriate hardware for particular solution. There are different types of hardware in the market such as integrated circuit–based hardware for running particular program after uploading the program and another type of hardware is based on operating system. Above two types have their advantages as well as limitation. Before purchasing the hardware, we should know about the capability of device and project requirement because power consumption and cost play vital role for any project. Following are some hardware devices with their merits and demerits.

TABLE 5.2   List of IoT Hardware with Their Merits and Demerits.

Sl. no.	Hardware	Merits	Demerits
1	Arduino UNO	• Low cost • Less power consumption • IC based • 14 digital pins where 6 provides PWM • 6 analog pins	• Network connectivity is not available
2	Arduino Mega 2560	• Low cost • Less power consumption • IC based • 54 digital pins where 15 provides PWM • 16 analog pins	• Network connectivity is not available
3	Bolt IOT	• Low cost • Bolt IoT cloud-based device	• 8 digital pins where 15 provides PWM • 4 analog pins
4	Raspberry Pi	• Operating system based • Single-board computer • Network connectivity • Camera, display, HDMI, and Ethernet slots • USB slots	• High cost • High power consumption
5	CC 3200	• Internet on chip • External serial flash boot loader • 32 I/O pins • Build in power management • Wi-Fi network processor • Onboard HTTP web server	• High cost

IoT devices are further classified into end device, communication module, and gateway modules. We can select appropriate modules as per requirement of our solution.

TABLE 5.3 IoT Hardware Modules for End Device, Communication and Gateway.

End devices	Commination module	Gateway module
Arduino UNO	Xbee S2C	Raspberry Pi
PIC 32 starter kit	ESP8266 NodeMCU	Intel_Galileo
MSP430G2	CC110L	CC3200

5.6 MAJOR APPLICATION OF IOT IN AGRICULTURE

Implementation of latest technology such as IoT in agriculture is changing the roots of traditional farming. Technologies are implementing on agro farms such as wireless technology for remote access and sensing technology for getting the environmental parameters and crop or soil conditions. Following are the practices that convert farming into smart farming and resolve farming issues and pest control. Some of the technological advancement helping in agriculture is discussed below.

5.6.1 SOIL MAPPING

Soil is the most important part for farming; so in first step, we are going to follow the soil measurement sensors. The main factor behind a good crop is the nutrition in the soil which can be determined by soil mapping. We can perform following tests over the soil using different sensors

1. Soil moisture
2. Nutrition in soil

This parameter directly reflects fertilization of the soil.

5.6.2 PEST CONTROL AND CROP DISEASE

Excess use of pest in the crop will affect the soil and human as well. So identifying the crop disease and using appropriate pest for disease control is possible using advance image processing technologies embedded with IoT systems.

5.6.3 IRRIGATION

Irrigation plays a vital role for any crop, and there should not be shortage of water as well as excess, otherwise it will affect the crop. About 97% of water on the Earth is salt-water held by oceans and only 3% of the remaining is fresh water. In different areas in the world, the fresh water availability is also different such as humidity in air, frozen water in the form of glaciers and underground water. Intelligent and smart farming using IoT will reduce excess of water used in agriculture and minimize the wastage. Using sensors we can find the requirement of the water and fulfill it.

5.6.4   FERTILIZER

Appropriate amount of fertilizer directly reflects the profit and healthy crop. Different soil monitoring sensors provide the status of the nutrition value in the soil, and as per the data, we can maintain the quality of soil and appropriate use of fertilizer.

5.6.5   CROP MONITORING AND HARVESTING

Advance image processing features in OpenCV library are implementing for crop monitoring and farmer will get the information and time for harvesting, which increases the quality of product and profit.

5.7   AGRICULTURE SENSORS, ACTUATORS, AND SOFTWARE

Above we have discussed a number of parameters that are useful in agriculture such as crop monitoring, harvesting, fertilizers, and irrigation. Now we will see the different sensors modules, actuators, and software that are useful in IoT for agriculture science.

TABLE 5.4   Different Sensors for Agriculture with Description.

Sl. no.	Sensor name	Image	Description
1	Soil Moisture Module		It is an easy-to-use soil moisture sensor and gives digital output of 5 V. It can measure water content or moisture level. The soil moisture sensor includes potentiometer to set threshold value [11].
2	DHT11 Humidity and Temperature Sensor		It is humidity and temperature sensor and gives digital output. It uses capacitive humidity sensor and thermistor to measure surrounding air.

TABLE 5.4 *(Continued)*

Sl. no.	Sensor name Image	Description
3	DS18B20 Waterproof Temperature Sensor	This is a sealed, waterproof, and wired digital temperature sensor.
4	Raspberry Pi camera board	This module is capable of 1080p video and image, connects directly with Pi board

TABLE 5.5 Different Actuators for Agriculture with Description.

Sl. no.	Actuator	Image	Description
1	Water flow controller		It controls flow of outgoing water using digital control.
2222	Thermoelectric module		It works on Peltier principle and converts electricity into temperature. Useful for indoor farming.

TABLE 5.6   Different Software/Platforms for IoT with Compatibility and Description.

Sl. no.	Software/platform	Compatibility	Description
1	Arduino IDE	Arduino microcontroller, NodeMCU	Integrated development environment for different microcontrollers
2	Python	Raspberry Pi	In operating system–based devices or in single-board controller, we use python as a programming language that gives web, cloud, and support different libraries.
3	Energia	CC3200 Launchpad	Integrated development environment for CC3200 Launchpad

5.8   AUTOMATIC WATERING SYSTEM FOR BASIL PLANT

Basil plant is a fragrant herb and used in Italian dishes. The most common type of basil is sweet basil and also there are other types such as purple basil, lemon basil, and Thai basil. This plant is easy to grow in outdoors. To plant basil, the top-most soil has to be warmed at least 10–20°C. This is a warm-weather plant.

To set up the smart watering system for basil we need following modules

1. Arduino Uno—The microcontroller and programmed using Arduino IDE
2. Soil moisture sensor—To get the soil moisture data
3. Jumper wires—To connect sensors and other relay with microcontroller
4. Water pump—For water supply
5. Relay—To trigger high-voltage electric appliances from low-voltage signal

Arduino Uno   Soil Moisture   Jumper wires   Water pump   Relay
                Sensor

FIGURE 5.4   Hardware modules for automatic watering system for basil plant.

Connection and Upload the code
1. Connect soil moisture sensors data pin to A0 of Arduino
2. Connect Relay modules signal pin to digital pin 4 of Arduino
3. Connect Ground and VCC pin of Sensor and Relay with Arduino
4. Make connection between water pump and Relay module (Fig. 5.5)
5. Complete the setup (Fig. 5.6)
6. Upload the following code in to Arduino from Arduino IDE.
   int sensorPin = A0; // soil moisture sensors data pin to A0 of Arduino

int relayPin= 4; // Relay modules signal pin to digital pin 4 of Arduino

int sensorValue;

int limit = 800; // Threshold value of soil moisture

void setup()

{

Serial.begin(9600);

pinMode(A0, INPUT);

pinMode(4, OUTPUT); }

void loop()   {

sensorValue = analogRead(sensorPin);

Serial.println("Analog Value : ");

Serial.println(sensorValue);

if (sensorValue<limit) {

digitalWrite(4, HIGH); } // Relay with get Signal Low and Water pump OFF

Else {

digitalWrite(4, LOW); // Relay with get Signal HIGH and Water pump ON

delay(5000); } // water pump is ON for 5 sec.

}

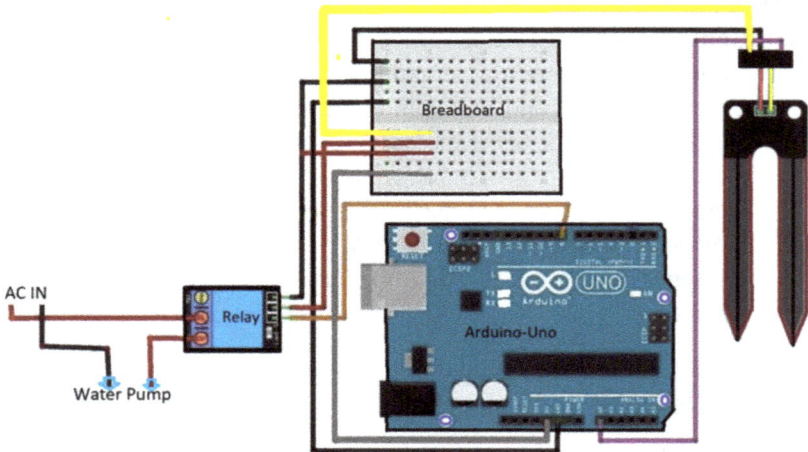

FIGURE 5.5   Connections between controller, soil moisture sensor, and relay.

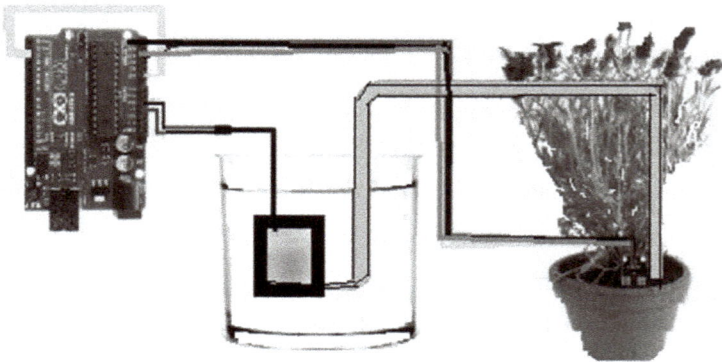

FIGURE 5.6   Automated plant-watering system.

5.9   ECONOMIC, ENVIRONMENTAL, AND SOCIAL IMPACT

IoT is an emerging technology for different types industries such as manu-facturing, maintenance, automobile, agriculture, and education. IoT refers to the machine-to-machine (M2M) interaction and communication that directly reflects the market. Nowadays agriculture and civilization both are demanding high production of food to fulfill the requirements of all over the world, and to make this possible, new technologies are being involving in the field of agriculture and providing efficient solutions.[1,3,6]

At the same time, the society is facing a new problem of water crises,[7] which is directly connected with agriculture; so to improve agricultural production, we have to resolve these problems as well. IoT is the technology that is being made easy to the farmers, so they can optimize the water uses and at the same time increase the production. In China, the introduction of IoT and cloud computing into agricultural method improvement and modernization is solving the problem. Key techniques of IoT and major features of cloud computing can build large data involved in agricultural production.[8] In India, up to 70% of the population directly and indirectly depends upon farming and the major part of capital income comes from agriculture. To overcome the shortcoming of agriculture, the only solution is smart farming, which is modernizing the current traditional farming methods.[9] Modernization of farming with IoT first provides remote controlling automation of task such as spraying, moisture–sensing, animal scaring, and keeping vigilance. Second, it includes intelligent decision-making and smart controlling based on real-time field data and, third, smart greenhouse management of different environmental parameters and theft detection.[9,10]

A number of factors such as climate change, heavy rainfall, heat, and intense storm results change the productivity of agriculture industry and directly affects the society. Smart agriculture with IoT based on sensor and cloud computing technology benefits the society in direct and indirect ways, which include optimization of energy resources, conservation of water, pollution prevention, and automation with low power consumption.[12]

5.10   IOT AND CLOUD PLATFORM FOR AGRICULTURE FRAMEWORK

The presented framework for agriculture is a six-layered concept. It includes hardware part as the first layer as physical layer having sensors actuators and microcontrollers. The second layer is network layer having the Internet and other network connectivity technologies.

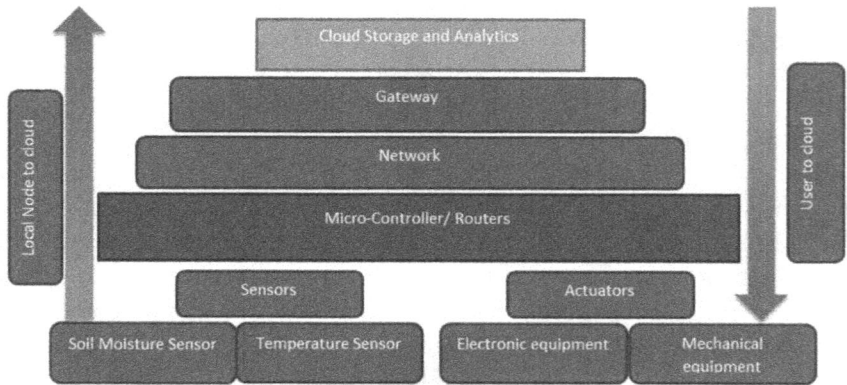

FIGURE 5.7   IoT and cloud platform for agriculture framework.

The third layer of framework is IoT middleware that performs context awareness, device management, platform portability, and security. The fourth layer is IoT cloud–aided service layer having vital role in providing software-as-a-service and cloud storage. Analytics layer is the sixth layer; this layer performs big-data processing and prediction analysis. The last layer of framework is user experience layer, the top-most layer that is designed for farmers.[13]

To implement IoT framework in efficient manner, we need to use appropriate cloud platform that can fulfill use requirements such as real-time data capturing, data visualization, data analytics, and the cost. In Table 5.6, we have cloud platform for IoT solutions and their respective features.[13,14]

TABLE 5.7   Cloud Platforms and There Features for IoT.

Cloud platforms	Cloud service type	Real-time data capture	Data visualization	Data analytics	Free account
ThingSpeak	Public	Yes	Yes	Yes	Yes
Temboo	Public	Yes	No	No	Yes
Ubidots	Public	Yes	Yes	Yes	Yes
Phytech	Private	Yes	Yes	Yes	Pay per use
Nimbits	Hybrid	Yes	Yes	No	Free

CONFLICTS OF INTEREST

The authors declare no conflict of interest.

KEYWORDS

- **agriculture**
- **Internet of things**
- **cloud platform**
- **IoT system design**
- **sensors**

REFERENCES

1. Hedley, C. B.; Knox, J. W.; Raine, S. R.; Smith, R. Water: Advanced Irrigation Technologies. In *Encyclopedia of Agriculture and Food Systems*; Academic Press: Cambridge, MA, 2014; pp 378–406.
2. Texas Instruments http://www.ti.com/technologies/internet-of-things/overview.html (accessed on 15 Dec 2019).
3. Burrell, J.; Brooke, T.; Beckwith Computing: Sensor Networks in Agricultural Production. *IEEE Pervasive Comput.* **2004,** *03* (1), 38–45.
4. Bandyopadhyay, S. K.; Sanyal, P. Application of Intelligent Techniques towards Improvement of Crop Productivity. *Int. J. Eng. Sci. Technol. (IJEST)* **2011,** 3 (1); ISSN: 0975-5462.
5. Lee, M; Nat, S. et al. In *Agricultural Production System Based on IoT,* Computational Science and Engineering , 2013 IEEE 16th International Conference, Dec 2013, pp 833–836.
6. Chen, N. et al. Integrated Open Geospatial Web Service Enabled Cyber-physical Information Infrastructure for Precision Agriculture Monitoring. *Comput. Electr. Agric.* **2015,** *111*, 78–91.
7. Jury, W. A.; Vaux, H. J., Jr. The Emerging Global Water Crisis: Managing Scarcity and Conflict between Water Users. *Adv. Agron.* **2007,** *95*, 1–76.
8. TongKe, F. Smart Agriculture Based on Cloud Computing and IoT. *J. Conv. Inf. Technol. (JCIT)* **2013,** *8* (2). doi : 10.4156/jcit.vol8.issue2.26
9. Gondchawar, N.; Kawirkar, R. S. IoT Based Smart Agriculture. *Int. J. Adv. Res. Comput. Commun. Eng.* **2016,** *5* (6). doi: 10.17148/IJARCCE.2016.56188.
10. Cai, K. Internet of Things Technology Applied in Field Information Monitoring. *Adv. Inf. Sci. Serv. Sci. AISS* **2012,** *4* (12), 405–414.

11. Wang, Q.; Terzis, A.; Szalay, A. A Novel Soil Measuring Wireless Sensor Network. *IEEE Trans. Instrum. Meas.* **2010,** 412–415.
12. Srisruthi, S.; Swarna, N.; Ros, G. M.; Elizabeth, E. In *Sustainable Agriculture Using Eco-Friendly and Energy Efficient Sensor Technology*, 2016 IEEE International Conference on Recent Trends in Electronics, Information & Communication Technology (RTEICT). doi:10.1109/rteict.2016.7808070
13. Ray, P. P. Internet of Things for Smart Agriculture: Technologies, Practices and Future Direction. *J. Ambient Intell. Smart Environ.* **2017,** *9*, 395–420. doi: 10.3233/AIS-170440
14. Zone, D.; Devada, A. Media Property, https://dzone.com/articles/internet-of-things-4-free-platforms-to-build-iot-p (accessed on 15 Dec 2019).

CHAPTER 6

Landslide Susceptibility Assessment Using a Low-Cost Wireless Sensor Network for Agricultural Land in Hilly Areas

SWAPNIL BAGWARI*, ANITA GEHLOT, and RAJESH SINGH

School of Electronics and Electrical Engineering, Lovely Professional University, Phagwara, Punjab, India

Corresponding author. E-mail: sbagwari@gmail.com

ABSTRACT

Growing population settlement nowadays has two big problems in terms of food supply and providing infrastructure space to make homes. Somehow we need to settle this issue in some extent for consumer occupying unstable, steep, or remote areas. Although to stabilize unstable or steep areas is too costly due to geographic conditions, they have no other option to relocate. Fortunately, with the enhancement in technology and low-cost devices or sensors, actions can be adopted to ensure an individual's safety or infrastructure of agricultural. It is strongly suggested that the government gives assistance to professional geologists' incorporations with engineers to make monitoring system for successful mitigation of unstable slopes and be consulted before actions are taken. Landslides caused watersheds in Guatemala after tropical storm in 2005, which affected farmers' livelihoods in terms of destroying seeds and food stocks and preventing access to farmlands. In this issue, a system is proposed using accelerometer sensor and wireless sensor network using Advance Virtual Reduced Instruction Set Computer (AVR) microcontroller-based computing unit to monitor potential chances of landslide in agriculture ground especially in hilly areas. The proposed system transmits data from one of the Xbee set as end node to another Xbee set as coordinator node. Same data is placed in the cloud to provide further processing such as machine learning and deep learning.

6.1 INTRODUCTION

Deployment of wireless senor in hilly areas is always a challenging task. Demographic characteristics of hilly areas are different; hence sensor placement is very important parameter to detect landslide chances in agriculture field. Factors such as connectivity links with Wi-Fi, network issue due to bad weather, and power issue are other issues that need to be taken care of. Real-time measurement has been performed with respect to network after deployment of wireless sensor network system (Kumar et al., 2019). Measurement of vibration was performed using accelerometer on MICAz devices and suggested that with changes of 0.2–0.49 g of either X or Y axis, the soil starts to move but not effectively. Changes of more than 0.5 g indicate changes in the ground (Kotta et al., 2011). Evaluation of different disasters such as fire, earthquake, flood, and landslide is performed in Uttarakhand. Various approaches are discussed using adopted sensors and wireless sensor network (Pant et al., 2017). An important factor that affects agriculture could be landslide. It is important to read the data of ground to study the landslide chances. Moreover the surrounding changes are also important to check such as temperature, humidity, and pressure. Wireless sensor network landslide monitoring system proposed to monitor slope angle and vibration to monitor hill areas. At rainy days, the moisture content extends by 20%, whereas in sunny days, it was only 3%. At this point, safety factor drops tremendously (Yunus et al., 2015). Sensor and GIS-based systems are designed to detect landslides in Darjeeling (RinaMaiti et al., 2018). Wireless sensor networks are meant to be deployed in very harsh conditions and may be for very long period of time. Here, quality of service should be maintained. Sensors failure, communication fault, and other unknown fault may change the topology. Hence fault detection is required to maintain the quality of service. Qualitative and quantitatively technique and algorithms are compared and discuss future directions to detect fault in wireless sensor networks (Muhammed et al., 2017). Artificial neural network and logistic regression are applied to datasets of remote sensing in the Jazan province, Saudi Arabia. Moreover more exercise is conducted such as data compilation of debris flow from satellite datasets and spatial analysis of geographical information system environment (Elkadiri et al., 2014). Monitoring system proposed for earthquake triggered landslide via WSN using star topology, Wi-Fi shield, GPRS, and Zigbee (Latupapua et al., 2018). Landslide detection system is developed for Bidholi, Dehradun, region where climate changes most of the time. Heterogeneous network created using geophysical sensors to identify fault and analysis of sensors data using wireless sensor network (Teja et al., 2014). A review of risk management for landslide is conducted with take care of location of landslide-prone area, placing of real-time system and analysis of continuous monitoring system (Suryawanshi et al., 2017). Wireless sensor network is created in different rooms to measure temperature by using Xbee in coordinator and end-node mode. Each device consist of microcontroller, Xbee works on IEEE standard 802.15.4. The coordinator has Ethernet shield to send data on web browsers (Boonnsawat et al., 2010). Two types of networks are compared for establishing the performance metrics in network throughput, received signal strength, mesh routing, packet transmission delay, and consumption of energy in indoor environment. First network created direct communication between coordinator and end device and in second network communication between end nodes to router node to coordinator node. Simulation has been performed in Docklight V2.0 software (Piyare et al., 2013). For Taziping landslide, a simulation platform is implemented using artificial neural network (Nastiti et al., 2016). Low-cost vibration sensor network is proposed using PIC12F683 microcontroller and Wi-Fi ESP8266 module. Vibration sensor is placed in stainless node using amplification circuit (Biansoongnern et al., 2016). In LuShan, Taiwan sensing system designed for smart soil particle to detect the tilts for potential landslide. Micro Electro Mechanical Systems based accelerometer and motion sensor is used with processing unit such as Raspberry Pi. The Processing unit is capable for executing large amount of data in less time (Ooi et al., 2016). In Honghui et al. (2017), displacement, angle, and rainfall sensor hardware architecture proposed for geohazards monitoring using Zigbee-based wireless sensor network. Authors proposed (EI Moulat et al., 2018) Lambda architecture to monitor landslides using monitoring of quality such as temperature, humidity, rainfall, soil moisture, and acceleration. Further data is sent to the cloud using ESP32 Chipset. In Ju et al. (2015), WebGIS-based information management system proposed to update the warning in mail and mobile message. Proposed system used Zigbee- and GPRS-based wireless sensor network to monitor sensor data. The authors (Nath et al., 2017) investigate the response of mesh routing and tree routing for twenty minutes for delay in topologies, load in coordinator, data collection and MAC load. Results show in case of failure and recovery procedure, network performs according to Zigbee standard. In Obaid et al. (2014), authors performed the survey of different wireless sensors presented and with it Zigbee in home automation application is also discussed. Authors (Chang et al., 2011) proposed monitoring of slope stability using tiltmeter- and Zigbee-based wireless sensor network. Data of different nodes are transmitted and received by management to take further decision. In Gili et al. (2000), performance of Global Positioning System (GPS) evaluated for precise measurement of coordinates. GPS allows productivity in all type of weather and large coverage even not requirement of on sight communication with base station. Result of GPS and fixed point sensors such as extensometers, EDM, and inclinometers are compared. In Ref. [28] machine-to-machine-based geological disaster monitoring system is proposed using GSM and Beiodu wireless network. Gao et al. (2020) proposed to get relationship between pest/disease and different parameters of weather. Due to remote areas and less infrastructure in hilly areas for power supply automatic sun tracking device is designed which will change the location of solar panel according to sun. In their system, unmanned aerial vehicles (UAVs) set in a flight mode that ensures maximum throughput and finally send the images to the cloud for further analyzing of damage by pest and diseases. In Anbalagan et al. (2008), the authors' proposed landslide hazard zonation mapping on the level of mesoscale, that is, 1:5000–10,000 gives guidance to choose proper area for urbanization and expansion in hill. Landslide hazard zonation (LHZ) map was applied on this scale in Nainital area of Uttarakhand. Sulaiman et al. (2017) designed accelerometer sensor network for railway tracking using Arduino. Response of reading data is less than 1 s and average error is 0–3.84%. Lee et al. (2010) monitored debris flow using multifunctional sensor networks. Sensors also drift with debris flow to get the information of movement along and transmit data wireless to base station. Kalaivani et al. (2011) performed survey on various applications of Zigbee-based wireless sensor network in the field of agriculture to check parameter such as weather, soil temperature, monitoring leaf, and weed disease.

6.2   TYPES OF LANDSLIDES

The term "landslide" describes a process that results in outward and downward movement of mass waste or mass movement of rock, soil, artificial fill, or combination of all. Mass movement may move by falling, sliding, flowing, spreading, or toppling. Efficient literature survey and research are available in hands to concern landslides, but unfortunately very less of it synthesized and integrated to address around the globe. Due to landslide, the extent of agriculture land is getting damaged, or in another scenario, due to lack of planning or knowledge, farmers start agricultural work in landslide-prone area. Land-use policies to choose agricultural areas in hills are outdated or not effectively used by farmers. Even nonexistent land-use policies cost local jurisdictions and may become nowadays state or national problem. Various types of mass movements are associated with term landslide. When different zones of weak soil detach from stable underlying material due to gravitational force, landslides occur.

There are two types of major slides that, rotational or translational slides, mostly affect hilly areas. In rotational slide, the top region of the crack bends upward, and the slide development will be rotational about a pivot corresponding to the ground surface and transverse over the slide, though in translational slide, the slide moves along organizer surface with less rotational or in reverse tilting.

Falls are abrupt movements such as bed rocks and boulders that get detached from cliffs of slopes. It is strongly influence by gravity or presence of interstitial water. This type of structure must be taken care of by farmers to avoid any mishap in the future. Topples is forward rotation of a unit exerted due to gravity in adjacent joints or by fluids in cracks. Flows are five basic categories and very important to monitor in regular bases for agriculture land. All five categories are different in their ways. Mostly flows are the major reason to destroy the infrastructure. Debris flow is a kind of fast mass development mix of free soil, rock, natural particles that stream downslope because of gravity. Debris flows are for the most part activated by serious water stream, quick snowmelt. It includes less than 50% fines. Trash torrential slide is an assortment of very quick debris flow. Earthflow and mudflow occur in fine-grained materials and wet materials, respectively. Creep type of flow is the imperceptibly slow due to shear stress. Seasonal creep mass movement based upon changes in soil moisture and temperature, whereas continuous creep failure totally depends upon shear stress.

6.2.1   RELATIONSHIP OF DIFFERENT FACTORS WITH LANDSLIDE

Following are the three major factors that are responsible for the chances of landslide:

6.2.1.1   LANDSLIDE AND WATER

Essential driver of landslide is slant immersion because of water consumption. This impact is firmly conceivable with precipitation and changes in levels of ground water, dams, waterways, etc. Debris flows and mudflows generally happened in regions that are small and in soaked channels because of floods, and these kinds of landslides happen all the while in a similar territory.

6.2.1.2   LANDSLIDES AND SEISMIC ACTIVITY

Earthquakes in steep landslide-prone area increase the monetary losses, destroying agricultural land in all possible way. It may allow rapid infiltration of water or make large crack. This must be monitored to reduce the chances of failure in the future. Cracks in sensitive area or steep can be monitored by camera processing. It is also important that ground shaking caused dilation of soil materials. Widespread rock falls caused loosening of rocks as a result of ground shaking.

6.2.1.3   LANDSLIDES AND VOLCANIC ACTIVITY

Landslide due to volcanic magma liquefy snow cause a deluge of rock, soil, debris, and water that expansion soak inclines which pulverizing anything in the way. The 1980 emission of Mount St. Helens in Washington set off a huge landslide on the north flank of the spring of gushing lava, the biggest landslide in recorded occasions.

6.3 SLOP STABILITY

In hilly areas for any aspect either for agriculture or construction questions needs to be addressed that how much the slop is stable. There will be always two forces applied, that is, resisting force and driving forces. Evaluation of relative magnitude of resisting forces and driving forces is required. Resisting force monitored stability, whereas driving forces monitored movement. If the driving force which is the pull of gravity on slope material is more than resisting forces then the chance of collapse or landslide is very prominent.

$$\text{Forces (N/m}^2) = \text{Thickness (m)} \times \text{density (kg/m}^3) \times 9.80 \text{ N/kg} \quad (6.1)$$

In a particular area, the total resisting forces will be the addition of normal force and cohesion. Change in cohesion, change in water content, and change in slop angle are the important parameters of driving forces. Factor of safety must be high, and value far to 1 will more stable, whereas value near to 1 will be sensitive for landslide.

$$\text{Factor of Safety} = \text{Resisting Forces/Driving Forces} \quad (6.2)$$

So it is very important to calculate factor of safety by government to provide suitable land to farmer for agriculture in mostly hilly areas. Wireless data also can be achieved in cost effective way by using different smart sensors and advance microcontrollers with long battery life.

6.4 PROPOSED SYSTEM

Proposed system comprises an accelerometer, ESP8266 Wi-Fi module, two Arduino Unos, and two Xbees. The system will detect the changes in ground using accelerometer and transmit the data from Xbee. Xbee and accelerometer will be connected with Arduino Uno and will be denoted as an end-node device. Xbee also needs to be set as end-node device using either by AT commands or XCTU software. Calculated data will be received by another Zigbee, which is set as coordinator using XCTU software. Once the data is received in gateway, further data is sent to the cloud using Wi-Fi module ESP8266. For Xbees, explorer board is used. Xbee required 3.3 V of power supply that is converted due to regulator IC present in explorer board as it has 5 V of power pin by default. In this way, there is no need to design a 3.3-V power supply channel using voltage divider circuit or regulator ICs. One of the Xbees connected with the accelerometer and soil moisture sensor. This Xbee is set as end device using XCTU software. Another Xbee is set as Coordinator device using XCTU software. End device will send the data in one-way communication to another coordinator Xbee that is connected with another Arduino Uno. In the proposed system, the accelerometer sends voltage signal and X, Y, and Z axis data to coordinator node. Both sensors are in the form of digital output sensor. Accelerometer sensor works in I2C serial communication protocol (Fig. 6.1).

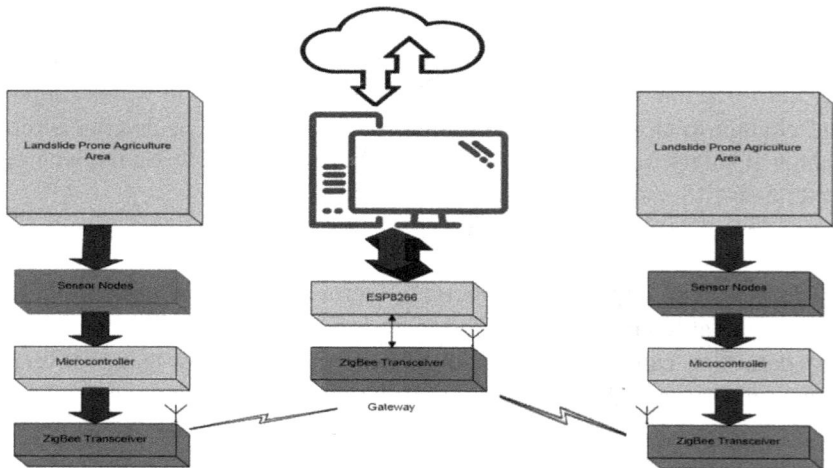

FIGURE 6.1   Block diagram of proposed system.

I2C stands for Inter-Integrated Circuit that uses only two pins to communication with Arduino Uno. One pin is called as SDA (serial data pin) and another is as SCL (serial clock pin). Another two pins are connected with ground and Vcc. Both Xbee connected with TX and RX pin of Arduino in cross-connection way, that is, Tx pin of Xbee is connected with Rx pin of Arduino and Rx pin of Xbee is connected with Tx pin of Arduino. Tx and Rx stand for transmit pin and receiver pin, respectively. In this example, power supply is provided by the computer but in real-time scenario, it can be provided by 9-V battery or solar panel. In that regard, voltage regulator ICs are required to maintain the power supply according to Arduino microcontroller, Xbee, and sensors. If the power is available from AC source then it must be converted to DC form by using diode and filtering circuit by using capacitor and inductor. L-type, Pi-type, or T-type filter circuit can be designed by capacitor and inductor. After filtering the DC signal, it must be converted to 5 V by using regulator IC, that is, 7805. Regulator IC diagram shown in the figure. For 3.3-V signal voltage, divider circuit can be designed using registers.

Following is the list of components used in proposed system to detect landslide in hilly areas. Xbee Explorer is not compulsory to use, without it by checking the pins, that is, transmitter (Tx), receiver (Rx), power supply (Vcc), and ground (Gnd) of Xbee system can be designed.

TABLE 6.1 Component list.

Component/specifications	Quantity
Xbee	2
Xbee Explorer	2
Arduino Uno	2
Accelerometer	1
Jumper Wire M-M	10
Jumper Wire M-F	10
ESP8266	1
ESP8266 adaptor board	1

6.4.1 ARDUINO UNO

In proposed system, high-performance, low-powered AVR 8-bit micro-controllers are used. It uses reduced instruction set computer architecture, which is better than complex instruction set of computer architecture. It is having 131 powerful instructions and most of the instructions executed using only one clock cycle. Because of high throughput, that is, 20 MIPS (million instructions per second), it is best choice to use in this project. Moreover data retention of microcontroller is about 20 years at 85°C and 100 years at 25°C. It has internal 16- and 8-bit timers/counters register with prescaler and compare mode. It also provides six pulse width modulation pins and six channels for analog pins, whereas digital pins are 14. It is also has good accuracy in wide range of temperature, that is, −40°C to 85°C.

AVR microcontroller have inbuilt features of power on reset, programmable brownout detection, and internal calibrated oscillator. Moreover, it
is having six modes of sleep, that is, power save, extended standby, ADC
noise reduction, idle mode, and power-down mode. It also supports Inter-
Integrated Circuit (I2C) and SPI (Serial Peripheral Interface) for serial
communication. As nowadays all other sensors and ICs such as memory
chips are based on I2C and SPI, it is very important that one's choose
microcontroller adequately. In terms of memory size, it has 32 kB of flash
memory, 1 kB of EEPROM (electrically erasable programmable read
only memory), and 2 kB of RAM (random access memory). It supports
also mechanism of read while writing self-programming method. Power
supply range is 1.8–5.5 V for 20-MHz crystal oscillator.

6.4.2 ACCELEROMETER

Accelerometer sensor node is used to measure tilt and vibration to landslide
monitoring. Three-axis directions can be monitored with respect to change
in velocity and magnitude. In the proposed system, ADXL345 accelerometer is used for their thin and small design. Moreover it is low-power
sensor with very high–resolution of 13-bit measurement. The accuracy
is around ±16 g. Output is terms in 16 bit and can be accessed by SPI or
Inter-Integrated Circuit (I²C). SPI module required either three- or four-
wire communication, and for I2C, two-wire communication is required.
It is having resolution of inclination to detect the changes as small as
0.25°C. It can work in wide range of temperature (−40°C to +85°C) and
have shock survival of 10,000 g.

The acceleration will be calculated using following equation

$$A = S_i/(1024 \times R - O_f)/S_s \qquad (6.3)$$

where A is the acceleration in all direction X, Y, and Z, S_i is the amount of
sampling, O_f is offset, R is the reference voltage, S_s is the sensitivity of the
accelerometer.

6.4.3 XBEE COMMUNICATION

Xbees available in different varieties depends on range, frequency, and
series. It is based upon IEEE 802.15.4 standard created by Zigbee alliance.

In Zigbee consist of different layer starting from application layer, API (application program interface) layer, security layer, networking layer, medium access (MAC), and physical radio (PHY). Application layer is controlled by customer, whereas API layer, security layer, and networking layer come under stack designed by Zigbee alliance. It can work in star, mesh, and cluster tree topology. States of nodes can be set as active or sleep. Xbee communicate with a serial port. XCTU software is required to see the changes in serial terminal.

TABLE 6.2   Pin diagram of Xbee module.

Signal	Pin	Type	Description
SCLK	28	I	Serial Peripheral Interface, maximum 8 MHz
SO	1	O	Serial output, Serial Peripheral Interface
SI	2	I	Serial input, Serial Peripheral Interface
CSn	3	I	Chip select, Serial Peripheral Interface
GPIO0	10	I/O	General Purpose Input Output Pin (Digital)
GPIO1	9	I/O	General Purpose Input Output Pin (Digital)
GPIO2	7	I/O	General Purpose Input Output Pin (Digital)
GPIO3	6	I/O	General Purpose Input Output Pin (Digital)
GPIO4	5	I/O	General Purpose Input Output Pin (Digital)
GPIO5	4	I/O	General Purpose Input Output Pin (Digital)
RESETn	25	I	Active low reset pin
VREG_EN	26	I	High signal to active voltage regulator
NC	15,18,21		Not connected

6.4.4   TYPES OF NODES

Zigbee is less-power, low-cost, less data rate, mesh network based on the IEEE 802.15.4 standard. It supports mesh capabilities of advanced routing.

6.4.4.1   COORDINATOR

Coordinator is the most capable node of Xbee topologies. It acts as a root of a network tree. There will be always one Xbee coordinator in any topology. Coordinator node is used to make the operating channel and personal area network for an entire network. Once the connection is achieved, the coordinator allows routers and end device to join. Further it can participate to route packets and act as source or destination for data packets.

6.4.4.2 ROUTER

Router maintains or creates network information, and by using this information, it will determine the best route for a data packet. A router must join network, so it can allow others routers and end device to join. This node will route data packets to other nodes or take data from other nodes. It is a mains powered node.

6.4.4.3 END DEVICE

End device cannot route information from any other device. This node will be either source or destination for data packets, but it cannot route packets. End devices are generally powered by battery and offer low-power operation.

6.5 XBEES AND X-CTU

The first step to communicate with Xbee is to select gateway between computer and Xbee. Xbee Explorer acts as a gateway between Xbee and computer. There are many options to choice as gateway such as Xbee Explorer USB, Xbee Explorer USB Dongle, and Xbee Explorer Serial. Xbee Explorer USB is mostly used as gateway and it requires USB cable to connect with computer. In this type of Explorer, mini-B USB port is provided. To translate data from computer to Xbee FT231X, USB to serial converter is used. There are four LEDs for Tx, Rx, RSSI, and Power. It also has a reset button and voltage regulator to provide sufficient amount of power supply. In Xbee Explorer, USB dongles have flexibility to connect it directly with computer or laptop. In this board, also a reset button, voltage regulator, and LEDs are provided. Xbee Explorer Serial will have bigger footprint in comparison to others Xbees Explorer. It also has same feature except that a switch is provided to turn it on or off. Lastly, after choosing explorer gateway, X-CTU software is required to install. This software is available for all type of operating system to provide serial monitor console.

6.5.1   ADDING XBEE

Two Xbees are connected in proposed system and shown as COM11 and COM12 in the below Figure 6.2. Both port set as baud rate to 9600, data bits 8, parity none, stop bit as 1, and flow control as none. At very first step after connecting with computer firmware is updated in both Xbees. Once the firmware is updated, AT (Attention) commands are executed to check both modules.

FIGURE 6.2   Adding two Xbees.

6.5.2   AT COMMANDS TESTING OF XBEE

In order to enter in AT commands, one needs to move from setting panel to serial monitor and connect. Right on the top of tool setting and serial monitor diagram as computer shown. Once the serial monitor options are opened, press the close link in both Xbees. After closing it, icon color converts to green which is a way to see that there is no error in connection and firmware.

To enter in AT commands, press in console three time plus(+++) sign. After successfully performed this step, press OK to return to Xbee. As in Figure 6.3, it is shown OK after it means Xbee is ready to take AT commands. In the figure, ATID is used to get personal area network ID, and ATSH is used to get upper half of Xbee address.

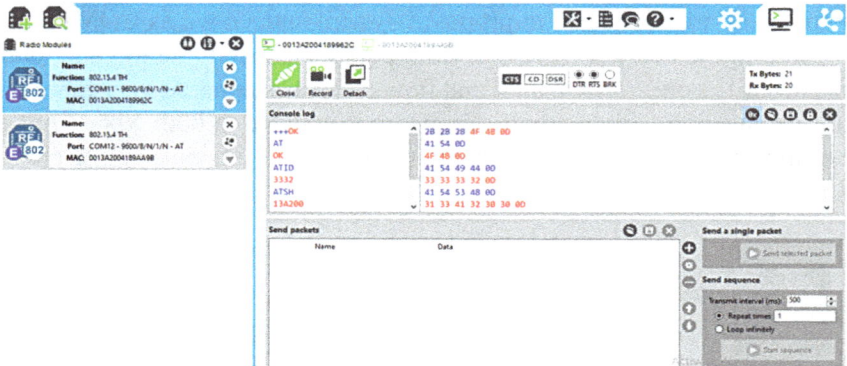

FIGURE 6.3   AT commands of Xbee.

6.5.3   TESTING CODE OF XBEE WITH ARDUINO

This code can be used to control external LED using Xbee. In this code, two options are provided; if the user will send Capital "A," it will turn ON external LED, and if user will send Capital "Z," then it will toggle LED 10 times with delay of 2 s in each duration of Toggle.

```
int led = 13; // Pin 13 as led
int received = 0; // Integer Variable set to zero
int i; // Integer variable
void setup() { // Initialize Pins and Baud rate
Serial.begin(9600); //  Baud rate set to 9600 bits per second
pinMode(led, OUTPUT); // To make led as output
}
void loop() { // Main code run infinitely
if (Serial.available() > 0) {// To check Serial data
received = Serial.read(); // To read Serial data and store it to received variable
if (received == 'A'){ // Compare received value with A
```

digitalWrite(led, HIGH); // If value is A is ASCII then On LED
delay(2000); // delay for 2 Seconds
digitalWrite(led, LOW); // Make LED Off after 2 Seconds
}
else if (received == 'Z'){ // if Value is equivalent to Z
for(i=0;i<10;i++){   // Loop for 10 Times
digitalWrite(led, HIGH); //This will make LED On
delay(2000); // delay for two Seconds
digitalWrite(led, LOW);// This will make LED Off
delay(2000); // delay for two Seconds
}}}}

6.5.4   COMMUNICATION BETWEEN TWO XBEEs

In order to communicate between two Xbees, certain changes are required in the modes of Xbee. It can be set as end device or coordinator. First, one needs to declare which node will be acted as end device and which one will be as coordinator device. End device will be have sensors connected and coordinator with route the traffic and receives the signal from end node. Accelerometer and end-node Xbee will be connected with the one of the Arduino Unos, whereas in the other Arduino coordinator, Xbee is connected (Fig. 6.4).

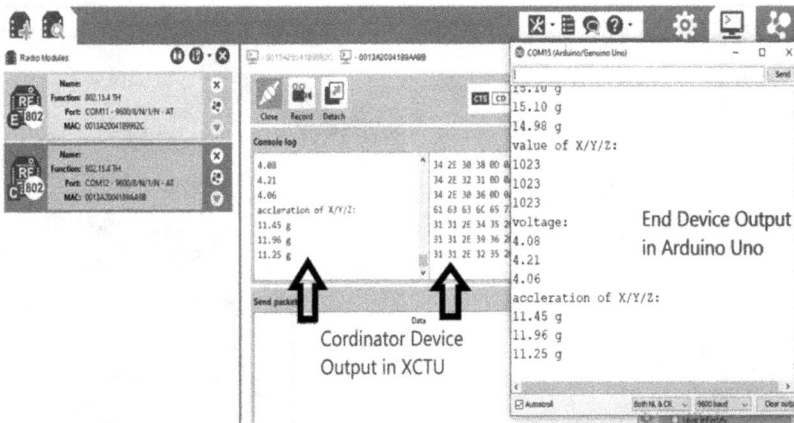

FIGURE 6.4   ADXL335 results between coordinator and end node.

To communicate between two Xbee, some settings are required under networking and security in tool. In first step, PAN, that is, personal area network, must be set uniquely for both modules. In this example, PAN is selected as 3332 for end-node Xbee and coordinator Xbee. For end-node device, DH destination address high is set to 0, DL destination address low is set to 12, and MY 16-bit source address is set to 10. Any values can be set between 0000H and FFFFH on these fields. For coordinator node, DH destination address high is set to 0, DL destination address low is set to 10, and MY 16-bit source address is set to 12. Any values can be set between 0000H and FFFFH on these fields. It needs to be taken care that DL and MY value will be inversely used in both the nodes (Figs. 6.5 and 6.6).

FIGURE 6.5   Hardware setup.

FIGURE 6.6   (a) Coordinator setup, (b) End node setup

6.5.5   PROGRAM FOR TRANSMITTER END

Following is the code of transmitter end that uses wire library to interface Inter-Integrated Circuit pins of accelerometer, adafruit_sensor.h library to resolve the issues any dependency of ADXL345 sensor library.

```
#include <Wire.h> // Wire Library
#include <Adafruit_Sensor.h> // Adafruit sensor library
#include <Adafruit_ADXL345_U.h> // Accelerometer Library
Adafruit_ADXL345_Unified accel = Adafruit_ADXL345_Unified();    // Function Call
void setup(void)
{
Serial.begin(9600); // Baud Rate Set to 9600 bits per second
if(!accel.begin()) //  Start of conversion signal
{
Serial.println("No ADXL345 sensor detected."); // Print Serial Data
}}
void loop(void) // put your main code here, to run repeatedly:
{
sensors_event_t event; // Read the event of sensor
accel.getEvent(&event);
Serial.print("X: ");  // Print Value of X in Serial Monitor
Serial.print(event.acceleration.x); // Serial Monitor Acceleration value of X
Serial.print(" ");
Serial.print("Y: "); // Print Value of Y in Serial Monitor
Serial.print(event.acceleration.y); // Serial Monitor Acceleration value of Y
Serial.print(" ");
Serial.print("Z: ");  // Print Value of  Z in Serial Monitor
Serial.print(event.acceleration.z); // Serial Monitor Acceleration value of Z
Serial.print(" ");
Serial.println("m/s^2 ");// Print in unit
delay(500);}
```

6.5.6   PROGRAM FOR RECEIVER END

Following is the code of receiver end, which uses serial receiving commands of Arduino IDE.

```
void setup() { // put your setup code here, to run once:
Serial.begin(9600); // Baud Rate Set to 9600 bits per second
Serial.print("Reciever End"); // Print the string in Serial Terminal
}
void loop() { // put your main code here, to run repeatedly:
if (Serial.available()>0){ // Wait for Serial Data
Serial.write(Serial.read()); // Write Serial Data on XCTU terminal
delay(2000); // Wait for 2 Seconds
Serial.print(Serial.read());// Print Serial Data on XCTU terminal
} }
```

6.5.7 PROGRAM TO UPLOAD SENSOR DATA IN CLOUD FROM GATEWAY

In this section, the data received by coordinator device is further sent to the cloud. There are many cloud services that are free such as ThingSpeak, Firebase, and AWS according to some extent. Here, for example, the services of ThingSpeak are used. ESP8266 Module designed by Espressif with adaptor is connected to microcontroller placed in gateway. ESP8266 is cost effective Wi-Fi module available with less space configuration. Module can be tested by running AT commands. In the program, some of the AT commands are used to connect with cloud such as AT+RST to reset Module, AT+CWJAP to connect with access point, AT+CIPSTART to connect with ThingSpeak cloud, and AT+CIPSEND to send data to particular IP. There are many other powerful Attention commands available which can be used in program according to the requirement of problem statement. Complete AT commands data sheet can be freely downloaded from the Internet. It is also very important to create account in ThingSpeak and get the write or read API key to either place data in cloud or read data from cloud. Once the APIs are provided, anyone can use the cloud services of ThingSpeak.

```
#include <Wire.h>// Wire Library
#include <Adafruit_Sensor.h> // Adafruit sensor library
#include <Adafruit_ADXL345_U.h> // Adafruit library of ADXL345
#include <SoftwareSerial.h> // Software Serial Library to make other pins
as Tx, Rx
```

```
String apiKey = "Write your Channel Key"; // Edit this CHANNEL API
key according to your Account
String Host_Name = "X XXXXXXXXX"; // Edit Host_Name
String Password = "Y YYYYYYYYY"; // Edit Password
SoftwareSerialser(3, 2); // RX, TX
Adafruit_ADXL345_Unified accel = Adafruit_ADXL345_Unified();   //
Function Call
void setup(void) // put code here to run repeatedly
{
Serial.begin(115200); // enable software serial
ser.begin(115200); // reset ESP8266
ser.println("AT+RST"); // Resetting ESP8266
char inv =""; // Format of CWJAP AT Command
String cmd = "AT+CWJAP";//To connect with Access Point
cmd+= "="; // Format of CWJAP AT Command
cmd+= inv; // Format of CWJAP AT Command
cmd+= Host_Name;//Host Name
cmd+= inv; // Format of CWJAP AT Command
cmd+= ","; // Format of CWJAP AT Command
cmd+= inv; // Format of CWJAP AT Command
cmd+= Password;//Host Password
cmd+= inv; // Format of CWJAP AT Command
ser.println(cmd); // Connecting ESP8266 to your WiFi Router
}
void loop(void) // put code here to run repeatedly
{
sensors_event_t event; // Function to read event
accel.getEvent(&event);
  /*
Serial.print("X: "); Serial.print(event.acceleration.x); Serial.print(" ");
Serial.print("Y: "); Serial.print(event.acceleration.y); Serial.print(" ");
Serial.print("Z: "); Serial.print(event.acceleration.z); Serial.print(" ");
Serial.println("m/s^2 ");
delay(500); // Delay for half seconds
  */
float x=event.acceleration.x; // To get the value of X variation
float y=event.acceleration.y; // To get the value of Y variation
float z=event.acceleration.z; // To get the value of Z variation
```

```
String  cmd  =  "AT+CIPSTART=\"TCP\",\"";  // Establishing  TCP
connection
cmd += "184.106.153.149"; // api.thingspeak.com
cmd += "\",80"; // port 80
ser.println(cmd);
Serial.println(cmd);
if(ser.find("Error")){
Serial.println("AT+CIPSTART error"); // Network Issues
return;
}
String getStr = "GET /update?api_key="; // prepare GET string
getStr += apiKey;
getStr +="&field1="; // Update first field data
getStr +=String (x);
getStr +="&field2="; //Update second field data
getStr +=String (y);
getStr +="&field3="; //Update second field data
getStr +=String (z);
getStr += "\r\n\r\n";
cmd = "AT+CIPSEND="; // Attention Command to send data to IP
cmd += String(getStr.length()); // Total Length of data
ser.println(cmd);
Serial.println(cmd);
if(ser.find(">")){
ser.print(getStr);
Serial.print(getStr);
  }
else{
ser.println("AT+CIPCLOSE"); // closing connection
Serial.println("AT+CIPCLOSE"); // Serial Print of AT command
}
delay(1000); // Update after every 1 seconds
}
```

6.6   CONCLUSION AND FUTURE SCOPE

Landslide is mostly a local event according to the geographical changes.
Agriculture can affect by landslide immensely if proper evaluation is not

performed either periodically or continues. If any changes occur and not monitored efficiently then the farmers can suffer immensely. For that reason, an attempt is created to get the knowledge of hilly areas by placing the accelerometer sensors and taking the real-time data using Xbees. Figures 6.7 and 6.8 shows the data updated in cloud and saved in locally in computer. In both ways, visualization is important to take decision in real time to avoid any unforeseen situation. It will be still be issue that data may not be interpreted by farmer directly, but giving proper education of threshold value problem can be solved in some extent. Moreover it can be achieved more efficiently if the alarm is auto-generated by machine by using machine learning or deep learning.

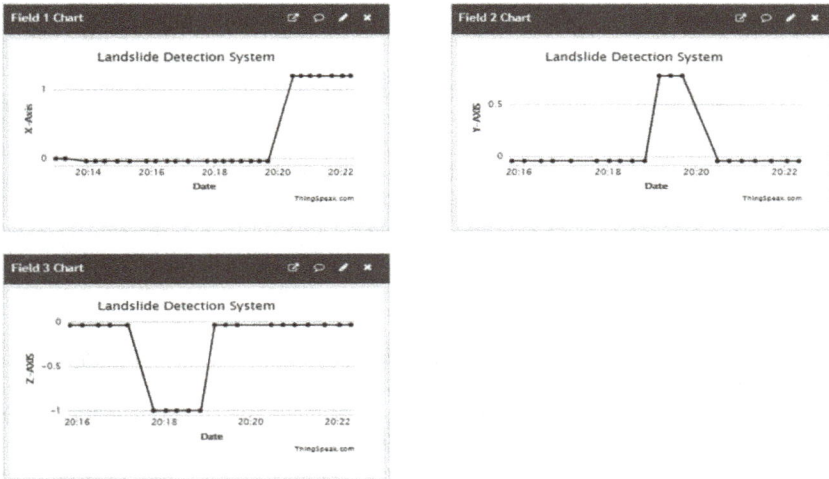

FIGURE 6.7 Data visualization in cloud.

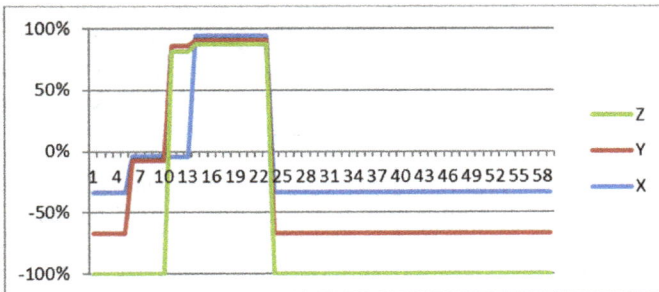

FIGURE 6.8 *X* axis, *Y* axis, and *Z* axis data of accelerometer.

Wireless sensor network is very emerging, not much costly technology and reliable to present real-time data. Further data can be analyzed by geological researcher to generate an alarm or provide education to farmer to use land for agriculture or not in hilly areas. It is also possible to apply machine-learning algorithms after consideration of geological data on the event area to generate auto alarm for farmers. This is one of the systems that can be useful to farmer and may to be motivated to design by government to help farmer in their agriculture area in hilly part of any country.

KEYWORDS

- **landslide**
- **wireless sensor network**
- **tri-axial accelerometer**
- **Xbee**
- **ESP8266**

REFERENCES

Anbalagan, R.; Chakraborty, D.; Kohli, A. *Landslide Hazard Zonation (LHZ) Mapping on Meso-scale for Systematic Town Planning in Mountainous Terrain*, 2008.

Arnhardt, C.; Asch, K.; Azzam, R.; Bill, R.; Fernandez-Steeger, T. M.; Homfeld, S. D.; Walter, K. Sensor Based Landslide Early Warning System-SLEWS. Development of a Geoservice Infrastructure as Basis for Early Warning Systems for Landslides by Integration of Real-time Sensors. *Geotechnol. Sci. Rep.* **2017,** *10,* 75–88.

Biansoongnern, S.; Plungkang, B.; Susuk, S. Development of Low Cost Vibration Sensor Network for Early Warning System of Landslides. *Energy Procedia* **2016,** *89,* 417–420.

Bizimana, H.; Sönmez, O. Landslide Occurrences in the Hilly Areas of Rwanda, Their Causes and Protection Measures. *Disaster Sci. Eng.* **2015,** *1* (1), 1–7.

Boonsawat, V.; Ekchamanonta, J.; Bumrungkhet, K.; Kittipiyakul, S. In *XBee Wireless Sensor Networks for Temperature Monitoring*, The Second Conference on Application Research and Development; Chon Buri, Thailand, 2010; pp 221–226.

Chang, D. T. T.; Guo, L. L.; Yang, K. C.; Tsai, Y. S. In *Study of Wireless Sensor Network (WSN) Using for Slope Stability Monitoring*, 2011 International Conference on Electric Technology and Civil Engineering (ICETCE); IEEE, 2011; pp 6877–6880.

Dan, L. I. U.; Xin, C.; Chongwei, H.; Liangliang, J. In *Intelligent Agriculture Greenhouse Environment Monitoring System Based on IOT Technology*, 2015 International

Conference on Intelligent Transportation, Big Data and Smart City; IEEE, 2015; pp 487–490.

Digi International Inc. XCTU; Next Generation Configuration Platform for XBee/RF Solutions. http://www.digi.com/products/xbeerf-solutions/xctu-software/xctu.

El Moulat, M.; Debauche, O.; Mahmoudi, S.; Brahim, L. A.; Manneback, P.; Lebeau, F. Monitoring System Using Internet of Things for Potential Landslides. *Procedia Comput. Sci.* **2018**, *134*, 26–34.

Elkadiri, R.; Sultan, M.; Youssef, A. M.; Elbayoumi, T.; Chase, R.; Bulkhi, A. B.; Al-Katheeri, M. M. A Remote Sensing-based Approach for Debris-flow Susceptibility Assessment Using Artificial Neural Networks and Logistic Regression Modeling. *IEEE J. Select. Topics Appl. Earth Observ. Remote Sens.* **2014**, *7* (12), 4818–4835.

Farahani, S. ZigBee *Wireless Networks and Transceivers: The Complete Guide for RF/ Wireless Engineers*; Newnes: Amsterdam (an imprint of Butterworth-Heinemann Ltd), 2018.

Gao, D.; Sun, Q.; Hu, B.; Zhang, S. A Framework for Agricultural Pest and Disease Monitoring Based on Internet-of-Things and Unmanned Aerial Vehicles. *Sensors* **2020**, *20* (5), 1487.

Gili, J. A.; Corominas, J.; Rius, J. Using Global Positioning System Techniques in Landslide Monitoring. *Eng. Geol.* **2000**, *55* (3), 167–192.

Honghui, W.; Xianguo, T.; Yan, L.; Qi, L.; Donglin, N.; Lingyu, M.; Jiaxin, Y. Research of the Hardware Architecture of the Geohazards Monitoring and Early Warning System Based on the IoT. *Procedia Comput. Sci.* **2017**, *107*, 111–116.

Huircán, J. I.; Muñoz, C.; Young, H.; Von Dossow, L.; Bustos, J.; Vivallo, G.; Toneatti, M. ZigBee-based Wireless Sensor Network Localization for Cattle Monitoring in Grazing Fields. *Comput. Electr. Agric.* **2010**, *74* (2), 258–264.

Ju, N. P.; Huang, J.; Huang, R. Q.; He, C. Y.; Li, Y. R. A Real-time Monitoring and Early Warning System for Landslides in Southwest China. *J. Mountain Sci.* **2015**, *12* (5), 1219–1228.

Kalaivani, T.; Allirani, A.; Priya, P. In *Survey on Zigbee Based Wireless Sensor Networks in Agriculture*, 3rd International Conference on Trendz in Information Sciences & Computing (TISC2011); IEEE, 2011; pp 85–89.

Kebaili, M. O.; Foughali, K.; FathAllah, K.; Frihida, A.; Ezzeddine, T.; Claramunt, C. Landsliding Early Warning Prototype Using Mongo DB and Web of Things Technologies. *Procedia Comput. Sci.* **2016**, *98*, 578–583.

Kotta, H. Z.; Rantelobo, K.; Tena, S.; Klau, G. Wireless Sensor Network for Landslide Monitoring in Nusa Tenggara Timur. *TELKOMNIKA Indonesian J. Electr. Eng.* **2011**, *9* (1), 9–18.

Kumar, S.; Duttagupta, S.; Rangan, V. P.; Ramesh, M. V. Reliable Network Connectivity in Wireless Sensor Networks for Remote Monitoring of Landslides. *Wireless Netw.* **2019**, 1–16.

Latupapua, H.; Latupapua, A. I.; Wahab, A.; Alaydrus, M. Wireless Sensor Network Design for Earthquake's and Landslide's Early Warnings. *Indonesian J. Electr. Eng. Comput. Sci.* **2018**, *11* (2), 437–445.

Lee, H. C.; Banerjee, A.; Fang, Y. M.; Lee, B. J.; King, C. T. Design of a Multifunctional Wireless Sensor for In-situ Monitoring of Debris Flows. *IEEE Trans. Instr. Measure.* **2010**, *59* (11), 2958–2967.

Muhammed, T.; Shaikh, R. A. An Analysis of Fault Detection Strategies in Wireless Sensor Networks. *J. Netw. Comput. Appl.* **2017**, *78*, 267–287.

Nastiti, H. T.; Praditya, I. E.; Mustika, I. W. In *Evaluation of XBee-Pro Transmission Range for Wireless Sensor Network's Node Under Forested Environments Based on Received Signal Strength Indicator (RSSI)*, 2016 2nd International Conference on Science and Technology-Computer (ICST); IEEE, 2016; pp 56–60.

Nath, S. K.; Aznabi, S.; Islam, N. T.; Faridi, A.; Qarony, W. Investigation and Performance Analysis of Some Implemented Features of the ZigBee Protocol and IEEE 802.15. 4 Mac Specification. *Int. J. Online Biomed. Eng. (iJOE)* **2017**, *13* (01), 14–32.

Obaid, T.; Rashed, H.; Abou-Elnour, A.; Rehan, M.; Saleh, M. M.; Tarique, M. ZigBee Technology and Its Application in Wireless Home Automation Systems: A Survey. *Int. J. Comput. Netw. Commun.* **2014**, *6* (4), 115.

Ooi, G. L.; Tan, P. S.; Lin, M. L.; Wang, K. L.; Zhang, Q.; Wang, Y. H. Near Real-time Landslide Monitoring with the Smart Soil Particles. *Japanese Geotech. Soc. Spl. Pub.* **2016**, *2* (28), 1031–1034.

Pant, D.; Verma, S.; Dhuliya, P. In *A Study on Disaster Detection and Management Using WSN in Himalayan Region of Uttarakhand*, 2017 3rd International Conference on Advances in Computing, Communication & Automation (ICACCA); IEEE (Fall), 2017; pp 1–6.

Piyare, R.; Lee, S. R. Performance Analysis of XBee ZB Module Based Wireless Sensor Networks. *Int. J. Sci. Eng. Res.* **2013**, *4* (4), 1615–1621.

Qiao, G.; Lu, P.; Scaioni, M.; Xu, S.; Tong, X.; Feng, T.; ... Li, R. Landslide Investigation with Remote Sensing and Sensor Network: From Susceptibility Mapping and Scaled-down Simulation Towards In Situ Sensor Network Design. *Remote Sens.* **2013**, *5* (9), 4319–4346.

RinaMaiti, S.; Mishra, D. L. GIS and Sensor Based Rain Water Harvesting with Artificial Intelligence System for Free Landsliding. *Int. J. Civil Eng. Technol.* **2018**, *9* (6).

Sulaiman, F.; Parinduri, I. H.; Abdullah, R.; Firmansyah, T.; Syarif, M. S. Accelerometer Sensor Applications Early Warning System Train Accidents Due to Landslide at Laboratory Scale. *IOP Conf. Ser.: Mater. Sci. Eng.* **2017**, *180* (1), 012152).

Suryawanshi, S. R.; Deshpande, U. L. In *Review of Risk Management for Landslide Forecasting, Monitoring and Prediction Using Wireless Sensors Network*, 2017 International Conference on Innovations in Information, Embedded and Communication Systems (ICIIECS); IEEE, 2017; pp 1–6.

Teja, G. R.; Harish, V. K. R.; Khan, D. N. M.; Krishna, R. B.; Singh, R.; Chaudhary, S. In *Land Slide Detection and Monitoring System Using Wireless Sensor Networks (WSN)*, 2014 IEEE International Advance Computing Conference (IACC); IEEE, 2014; pp 149–154.

Yunus, M. A. M.; Ibrahim, S.; Khairi, M. T. M.; Faramarzi, M. The Application of WiFi-based Wireless Sensor Network (WSN) in Hill Slope Condition Monitoring. *JurnalTeknologi* **2015**, *73* (3), 75–84.

CHAPTER 7

Crop Survival Analysis Using IoT and Machine Learning

CH SAI DINESH REDDY*, SWAPNIL BAGWARI, ANITA GEHLOT, AND RAJESH SINGH

Lovely Professional University, Phagwara, Jalandhar 144411, India

*Corresponding author. E-mail: chsdr24@gmail.com

ABSTRACT

The application of IoT in agriculture has proven its advantage by predicting the survival of crops in selected types of soil. Most of the people rely on agriculture in India. Nowadays technology is used to get accurate results in the area of agriculture. Moreover present technology, soil plays a most crucial impact in the area of agriculture. Agriculture is one of the applications of IoT. Varieties of soil are used to cultivate different types of crops. Each type of soil has its features and characteristics. To predict the survival of crops in selected soil machine learning techniques are used. In this study, a model is proposed that will predict the accuracy of the survival of Sugarcane crop in a Loamy type of soil by extracting the data from the soil with the use of smart sensors. Machine learning methods such as K-Nearest Neighbor's (k-NN), Random Forest and Support Vector Machine (SVM) models are used to predict the survival accuracy. After obtaining the results, it shows that k-NN performs better than the other two models.

7.1 INTRODUCTION

Most of the information about the work is shared via the internet. The concept of linking objects and devices using the internet to share or store the data among devices is known as IoT. The Internet was founded in the year 1983. Before the invention of internet, mobile phones were being used to communicate with people who are far away from us. There was a drastic change in the development of technology and the internet was one of the things which provide information about various topics that can be accessed with the help of the World Wide Web (WWW). In 2004, electronic mail (E-Mail) was introduced which can share the information with some other email addresses with the help of the internet. The use of the internet had been increased and help in providing various services like E-Commerce, online shopping, etc. Nowadays social media has a major role in human life (Fig. 7.1).

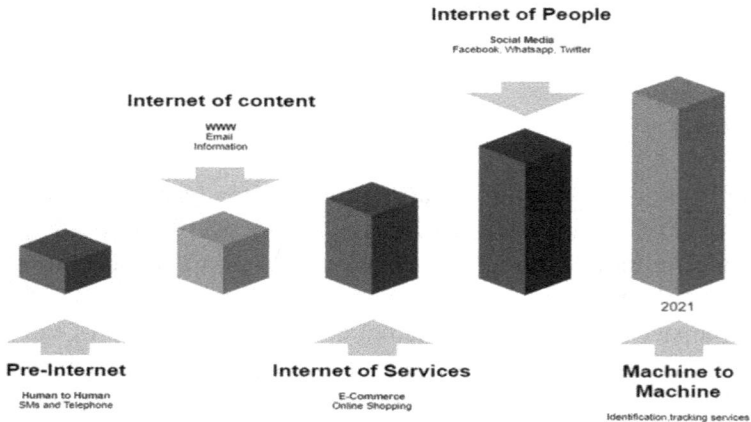

FIGURE 7.1   Overview of IoT.

With the help of the internet, it can provide some services like chatting, video calling, group chats, etc. Kevin Ashton proposed a concept to link RFID objects with the internet. At present, we were focusing on Machine-to-Machine communication typically known as M2M. This type of communication will help the objects to take decisions through some sort of training. This helps humans work easily. IoT is a combination of both devices and services that are responsible for collecting information, visualizing data, remote control, or smart controlling capability. For example, the best example is controlling a night lamp or bulb using an IoT device typically Amazon Alexa. Alexa is an artificially intelligent device that can control the IoT devices like night lamp to turn it on or off. IoT reduces the manpower or burden by making things smart. The Internet is required to control any smart device. In healthcare industries, nearly 76% of the products were being used followed by 67 and 66% by transportation and manufacturing.

Zhao et al.[1] proposed that IoT has a great impact on an important role in the field of agriculture. Separate sensors are used to measure humidity and temperature values. An analog to digital converter is also used to convert the analog signals into digital. This leads to an increase in hardware complexity and a lot of wiring. The better way is to use DHT 11 module which can measure temperature and humidity values simultaneously. Patil et al.[2] state the Internet of things technology reviewed in the available systems of the agriculture system. Refine the objective of this research study was to acquire real-time data related to agricultural production. Ubimote was used to collect and transmit the data. Pallavi et al.[3] proposed remote sensing of agricultural data and control system to the greenhouse agriculture. MAD architecture is used to upload the values to the cloud and Labor Monitoring Device (LMD) was used to monitor the labor work in the field. Labor will be given a wrist elastic band wherein an exceptional RFID number will be inserted. With the assistance of the gauging machine and computational unit, the number of yields gathered by the work will be determined. This data will be transferred to the cloud by the LMD. Bing et al.[4] had explained how technology helps in the agricultural sector. The shrewd framework was created to control the yield development condition and to upgrade natural product planting the board. If the framework is planted in a lot of territories, it will be helpful for the higher measures of creation. A specialist framework is planned right now helps keep up the yield. Davcev et al.[5] were interested in using a low power consumption, long-range data communication device called LoRaWAN. This device can able to transmit the data over large distances. The data from the sensor nodes are collected and sent to the LoRaWAN base stations. The data collected that are sent to The Things Network (TTN) which is a platform to collect and store information from different LoRa devices. Math et al.[6] state that IoT based real-time local weather station for PA that would provide farmers a means of acting on their agricultural practices. Sensors like pH, soil moisture, DHT 11, light sensor are used to measure the values of the soil and those values are visualized in a cloud platform. This paper provides full support for the development of an IoT architecture. Rahman et al.[7] proposed that a model that can foresee soil arrangement with land type and as per forecast it can propose appropriate yields. A few machine learning calculations, for example, K-Nearest Neighbor (k-NN), Bagged Trees, and kernel-based SVM are utilized for soil classification. Many datasets have been filtered and the labels which are necessary for the classification are extracted. The results that they get show that SVM performs good accuracy when compared to k-NN and Bagged Trees. Providing fertilizer recommendation is the future concern of the researcher. Chilingaryan et al.[8] reasons that the fast advances in detecting advances and ML procedures will give financially savvy and complete answers for better yield and condition state estimation and dynamic. RS-based methodologies require the handling of colossal measures of remotely detected information from various stages and, in this manner, more prominent consideration is as of now being dedicated to AI (ML) strategies. This paper talks about research improvements directed inside the most recent 15 years on AI-based procedures for exact harvest yield expectation and nitrogen status estimation. Prakash et al.[9] clarifies how machine learning strategies, for example, multiple linear regression, SVM and recurrent neural systems for a forecast of soil dampness for 1 day, 2 days, and 7 days ahead. In this paper, the datasets are collected from different regions and the Mean Square Error (MSE) and Coefficient of determination (R^2). A good accurate result was given by multiple linear regression models. Goldstein et al.[10] conveys that the data from the sensors is sometimes noisy. To use the data by avoiding the noise, the author proposes that considering the moisture values smaller than 17 or higher than 39% is abnormal, and was filtered out. The performance of the simple linear regression models for the different datasets is calculated and shown in the table. Linear regression and two nonparametric approaches, GBRT and BTC are used to determine the models. The models were trained against eight different feature subsets and the result or the accuracy is shown. Charoen-Ung et al.[11] present a machine learning-based model for anticipating the sugarcane yield of an individual plot. The highlights utilized in the forecast comprise of the plot qualities (soil type, plot territory, groove width, plot yield/yield evaluation of the most recent year), Sugarcane attributes (stick class and type), plot development conspire (water asset type, water system technique, pandemic control strategy, fertilizer type/recipe) and rain volume. This paper provides information about the amount of yield per year in terms of accuracy and it also provides information that how they have applied machine learning and how they pre-processed the data. Their way of approach is good and can be helpful to predict the accuracy of the survival of sugarcane crops. They also contributed his work by developing a machine learning-based prediction algorithm like gradient boosting and random forest models for the yield prediction which can predict more accurately than human experts. Medar et al.[12] the authors defined how Remote Sensing Data and Machine Learning techniques have focused on the prediction of the crop yield. This method is cost-effective when compared to ground data. This paper helps me in finding the main difference in predicted algorithms for the survival of the crops based on different parameters.

Khanna et al.[23] state that Lora technology is used for long-range communication and can cover in rural areas. Ghadge et al.[30] explained how soil plays an important role in the field of agriculture. The system used two algorithms and the best will be chosen based on the output and thereby helping the farmers by reducing the difficulties. Priya et al.[18] proposed that farmers can attain good accuracy for the prediction of crop yield using the Random Forest classifier. They also explained how data mining plays a key role in the field of agriculture. Vanitha et al.[20] explained how to organize the dataset acquired from the field accurately. This also explains how to select the best features like rainfall in the dataset which will show a great impact on the results and how to build a model using it and alerting the farmer when needed. Reshma et al.[23] explained how IoT and machine learning helps the farmers in increasing productivity by using the available land without wastage and how to control the pests.

7.2   OBJECTIVES OF THE INTERNET OF THINGS

The main objective of IoT is to link every device with a common IP. This can be possible if the IPv4 was changed to IPv6, multi-protocol integration possibility, mobility, interoperability, and distributing its intelligence among various smart devices.

7.2.1   *COMMUNICATION TECHNOLOGIES*

LoRaWAN, 3G, 4G, 5G, Li-Fi, Wi-Fi, M2M, NFC, RFID are some of the examples of communication technologies that support IoT devices in terms of range, power consumption, radiation, availability, upgradability, and scalability (Fig. 7.2).

7.3 OVERVIEW OF SOIL

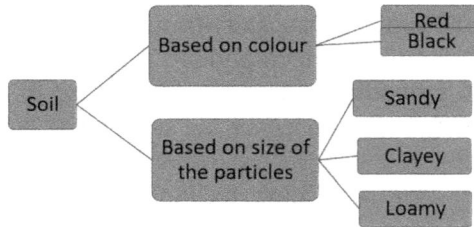

Soil
Based on colour — Red / Black
Based on size of the particles — Sandy / Clayey / Loamy

FIGURE 7.2   Soil types.

7.3.1 *BASED ON COLOR*

7.3.1.1 *RED SOIL*

The red color is due to the presence of iron oxide. This type of soil contains a mixture of sand and clay. This soil can be used by adding fertilizers and manures. This soil supports growing pulses, tobacco, cotton, etc.

7.3.1.2 *BLACK SOIL*

It is also known as black lava soil due to the formation of lava rocks. It has a high amount of clay. This type of soil supports growing millets, sugarcane, wheat, and oilseeds.

7.3.2 *BASED ON THE SIZE OF THE PARTICLES*

7.3.2.1 *SANDY SOIL*

It is porous and contains 60% of sand and clay. The size of soil particles ranges from 0.2 to 2.0 mm. There is a lot of air present in this soil due to poor water building capacity. This soil supports cultivating crops like melon and coconut. A large amount of water is required.

7.3.2.2 CLAYEY SOIL

The water keeping capacity of this type of soil is very high. The size of the particles is less than 0.2 mm. The organic capability of this soil is in the rich category. This soil supports cultivating crops like paddy which requires a high amount of water.

7.3.2.3 LOAMY SOIL

It consists of sand, clay, and silt. It has good water holding capacity with good aeration. Plant roots will get enough water, air, and space to grow. This soil supports cultivating crops like Sugarcane, Cotton, Jute, Pulses, and Oilseeds.

7.4 OVERVIEW OF SUGARCANE

Maharashtra, Karnataka, Tamil Nadu, and Andhra Pradesh hold the record of major sugarcane production regions in India. Sugarcane productivity is more in tropical regions. Seven states in India come under tropical region and nine states come under the subtropical region. The temperature conditions required for the growth of sugarcane are given as follows (Table 7.1).

TABLE 7.1 Temperature & Humidity Values of Sugarcane Crop.

S. no.	Sugarcane crop stages	Growth stages duration (days)	Temperature Max–min	Humidity Max–min	Sunshine in hours
1.	Germination	15–32	32–18	75–50	10
2.	Stem elongation	33–125	34–18	75–55	10
3.	Growth stage	125–210	34–22	85–75	11
4.	Ripening stage	211–365	15–12	65–40	10

The above table was taken from the ICAR-Indian Institute of Sugarcane Research which was printed in March 2017.

7.5 INTRODUCTION TO MACHINE LEARNING

Machine learning is a part of artificial intelligence. The goal of ML models is to predict, solve, make decisions, analyze, and fit the data into models.

It is a field in computer science but different from conventional algorithms that are programed. Machine learning has good decision making and can be trained with input data and desired outputs (Fig. 7.3).

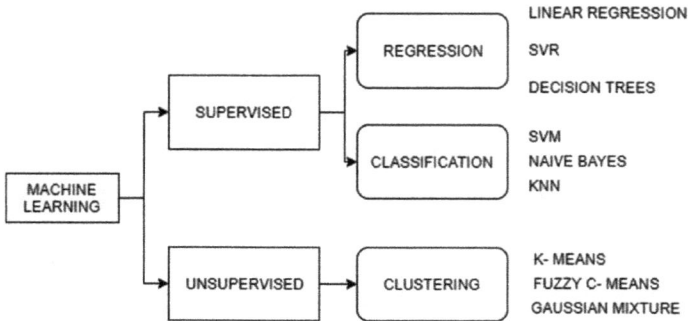

FIGURE 7.3   Methods of machine learning.

This research covers how IoT and machine learning play a vital role in precision agriculture. This project will explain how machine learning predicts the survival of the crop in soil based on the data acquired from the IoT platform. Machine mearning is not limited to image or face recognition, but it also provides good stand-in supporting the agricultural industry. In the same way, IoT is not limited to the Industry that deals with electronic components; it is also used for sharing the data related to soil among devices. Information from the IoT gadget is sent to the cloud. The cloud information is sent out and taken care of various calculations and the calculation gives exactness of a system. This research does not cover much about the Sugarcane and its varieties. It is purely based on the prediction of different soil irrespective of the type of crop. The objective of this study is to show how technology plays a key to support the agricultural industry by saving time, high productivity and yield, reducing manpower, reducing the usage of resources like fertilizers, and insecticides thereby there will be a decrease in the investment of crop. Farmers, as well as data scientists, will be benefitted from this kind of approach.

7.6   RESEARCH METHODOLOGY

In this project, the hardware is designed which resembles a field containing soil in it. A plot was designed to move across the field with the help of stepper motors. The sensors are attached to the plot or stand that will move around the field. The plot will move in a pattern that was preprogramed by the user. The sensors are made to contact the soil and the sensors begin detecting the properties of the soil and the information is sent to the micro-controller for the preparation. ThingSpeak is a cloud platform that will save the values sent by the NodeMCU. The information can be retrieved in CSV design whenever. The information which is acquired from the cloud is given to the ML models to foresee the endurance of the yield. The data mining tool called Jupyter Notebook used to predict the accuracy of the survival of the crop. The algorithms will process the data accordingly and give accuracy for the survival of the crop. By doing the above process the farmers will able to understand the real-time conditions of the soil and help in taking necessary action (Fig. 7.4).

7.6.1   FLOW CHART

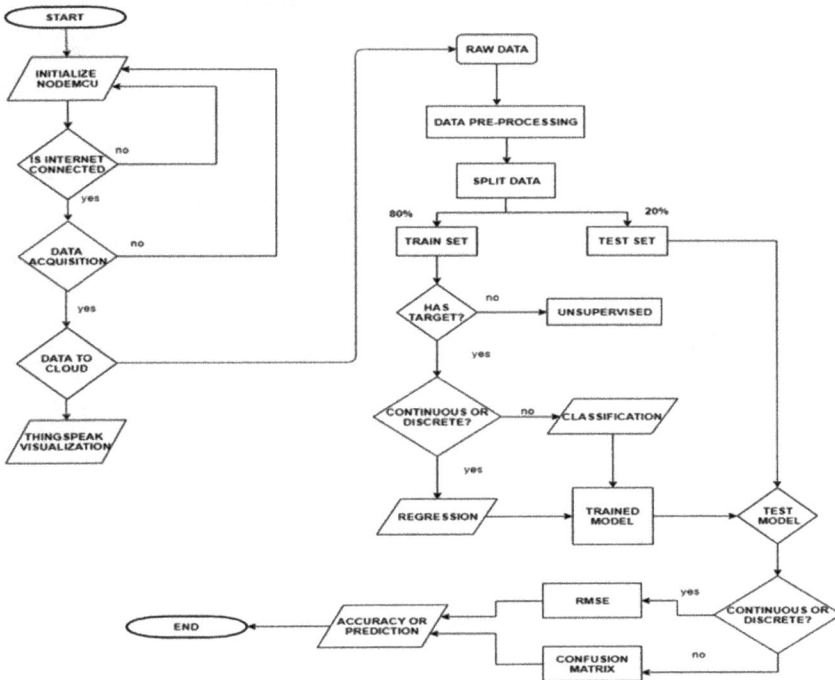

FIGURE 7.4   Flow chart.

The above flow chart represents how data are collected from the field with the help of sensors and sent to the cloud. First, NodeMCU is initialized and connected to a network. Sensors like DHT 11, Soil moisture, Rainfall sensor, and LDR module are connected and the data are sent to the cloud platform. ThingSpeak is an example of a cloud platform that is used to store and visualize the data to the user. Machine learning models such as SVM, k-NN, and Random Forest are implemented with the help of a programing language called Python.

7.6.2   WORK PLAN

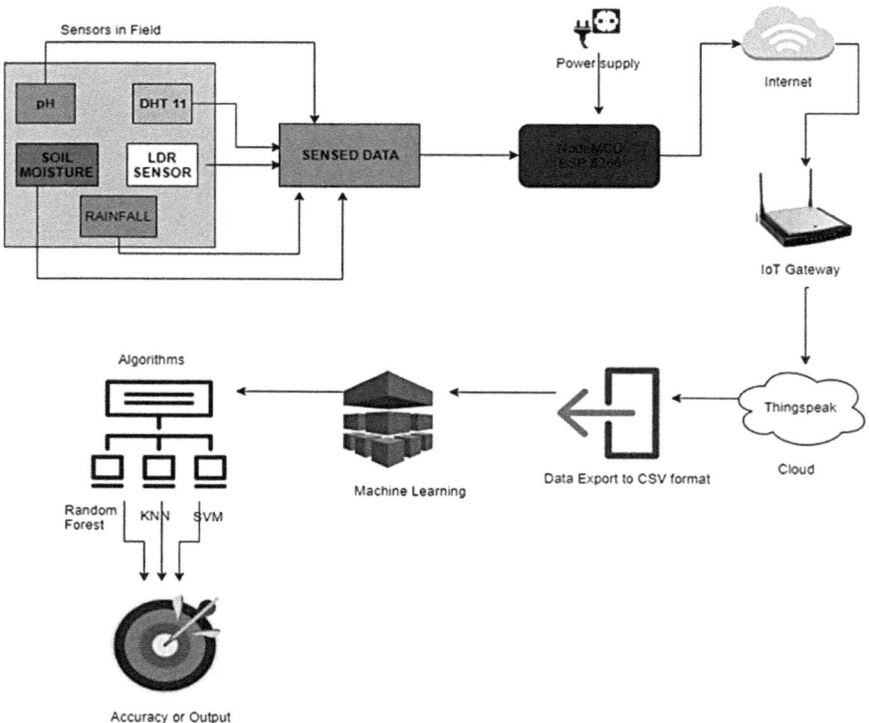

FIGURE 7.5   Proposed architecture.

Sensors like DHT 11, Rainfall sensor module, Light sensor module, pH sensor, and Soil moisture sensors are set in the soil (Fig. 7.5). These sensors are associated with the IoT gadget called NodeMCU and the gadget contains a Wi-Fi chip on it which is liable for sending the information to the cloud. The data were sent to the cloud platform called ThingSpeak. We can able to monitor the data in graphical form as well as numeric values. The data obtained are extracted to CSV format which will be accepted easily by the machine learning algorithms. The data are fed to the algorithms (RF, SVM, *and* k-NN) and the accuracy of the survival of crops will be predicted.

7.6.3 CIRCUIT DIAGRAM

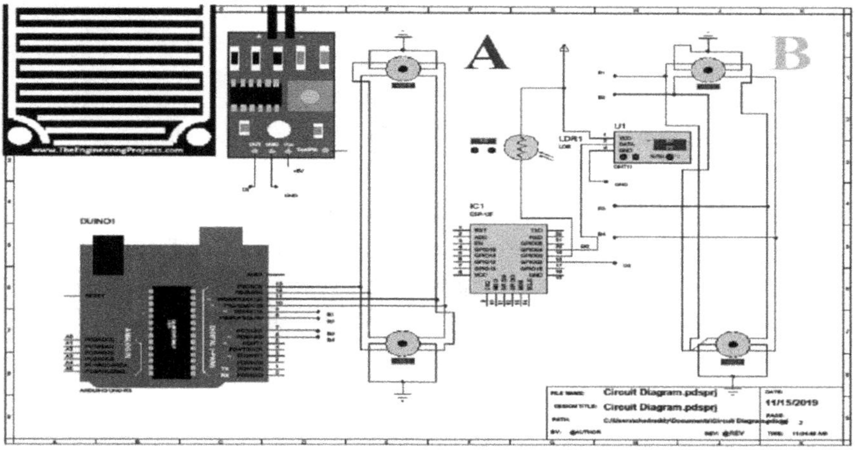

FIGURE 7.6   Circuit diagram.

The above figure explains how the sensors and stepper motors are connected to the microcontrollers (Fig. 7.6). On the A-side, two stepper motors are made in common and connected to the Arduino pins 13, 12, 11, 10 and on the other side of B two stepper motors are connected to the pins 9, 8, 7, 6. The motors will move up and down while the motors on side B will move from left to right. On top of the plot, NodeMCU is connected with sensors like DHT11, LDR, Rainfall sensor, etc. The NodeMCU will move throughout the field with the help of four stepper motors and collect the data using the sensors and sent it to the cloud.

7.7   EQUIPMENT, MATERIALS, AND EXPERIMENTAL SETUP

7.7.1   DESCRIPTION OF COMPONENTS

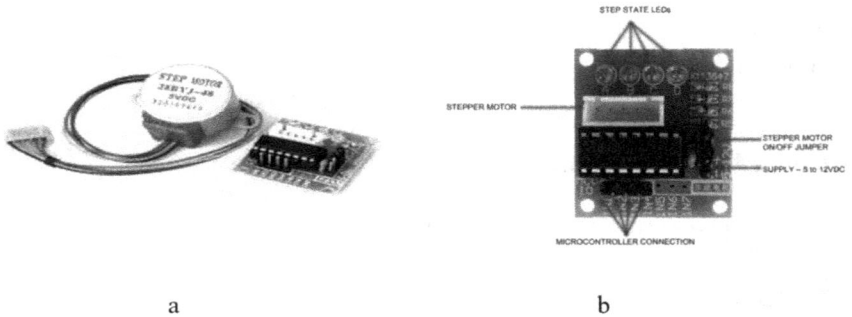

a   b

FIGURE 7.7   (a) 28BYJ-48 Stepper motor (b) ULN2003 Driver board.

Figure 7.7a is a unipolar device with four-phase five-wire that supports high precision with half step mode and high torque at full-step mode. Figure 7.7b, this board is required for the stepper motor to provide sufficient voltage along with the experiment. It can support up to 5–12 V DC (Fig. 7.8).

FIGURE 7.8   Timing belt.

In this project, the timing belt of the 2 cm width was used to connect the motor with pulley (Fig. 7.9).

FIGURE 7.9 DHT 11 module.

DHT 11 sensor module was utilized to figure the temperature and relative humidity of the soil. The operating voltage is about 3.5–5 V. The temperature range is 0–50°C and Humidity ranges from 20 to 90%. The module comes with a filtering capacitor and an inbuilt pull-up resistor (Fig. 7.10).

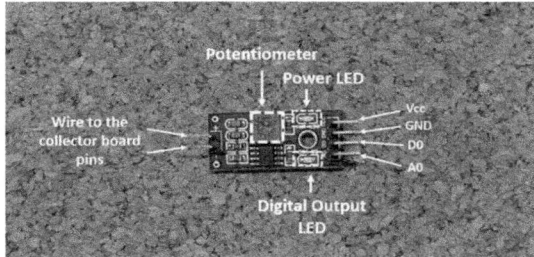

FIGURE 7.10 Soil moisture sensor.

The sensor that is shown in the above figure is used to measure the moisture percentage present in the soil. The YL-69 electrode is dipped into the soil and the data are transferred to the microcontroller. If the soil is dry then the output voltage increases and vice-versa. If the output is 30 then the soil is in the dry state if the value is between 30 and 70, it is humid soil and if the value is between 70 and 99, the soil is in water (Fig. 7.11).

a   b

FIGURE 7.11   (a) Light sensor module (b) Rainfall sensor.

This module contains an LDR sensor on top of it which is used to measure the intensity of the light. It will give output in the form of analog ranging from 0 to 1023 and digital ranging from 0 to 1. In this project, I mapped the values from 0 to 1023 to 0 to 50. In Figure 7.6b, this module will give information about the rainfall. If there is rainfall, it will reflect "1" otherwise "0."

a   b

FIGURE 7.12   (a) Arduino uno (b) NodeMCU.

In Figure 7.12a, Arduino is used to running the stepper motors and maintained sync between them. In Figure 7.12b, the data that are acquired from the sensors are subjected to the cloud using NodeMCU. This is an IoT device that contains the ESP 8266 Wi-Fi module embedded in it. It will use the MQTT protocol which acts as a broker between devices. This device will collect and transmit the information to the cloud platform.

7.7.2 SOFTWARE REQUIRED

7.7.2.1 ARDUINO IDE AND PYTHON

Arduino IDE is a user-friendly software application. This IDE is used to write the code for Arduino Uno and NodeMCU. The Arduino Uno code contains code written for driving the stepper motors accordingly and for NodeMCU, the code was written to send the data to the cloud platform with the help on the internet. The MQTT protocol acts as a broker between the NodeMCU and the cloud platform. Python is a general-purpose object-oriented language. In this project, Python is used to build different Machine Learning Models.

7.7.3 EXPERIMENTAL SETUP

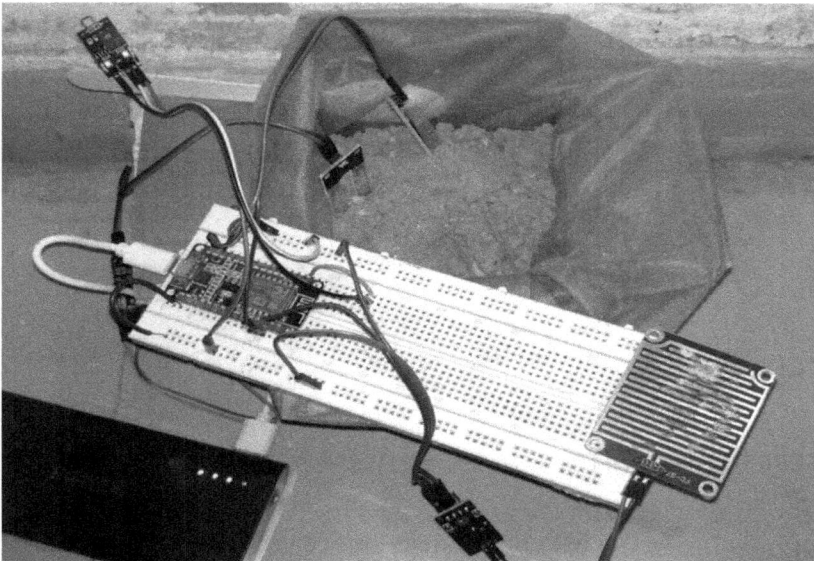

FIGURE 7.13   Experimental setup.

The above prototype was designed to explain the task of the system (Fig. 7.13). The setup plays a good role in the field of precision agriculture by collecting information and sorting using ML models. The data from the sensors are collected during October–November 2019. The below diagrams represent the ThingSpeak visualizations about various properties of the soil.

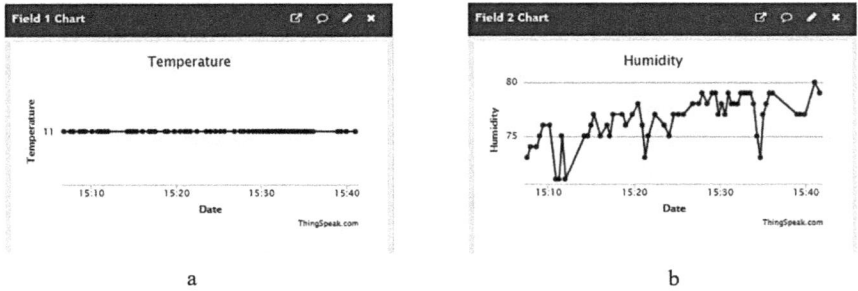

a   b

FIGURE 7.14   (a) Real-time temperature (b) Real-time humidity data.

Figure 7.14a shows the real-time temperature data sent by the NodeMCU using DHT 11. The variation in the humidity and temperature will show a great effect on the accuracy of the machine learning models. Figure 7.14b shows the real-time humidity data sent by the NodeMCU using DHT 11.

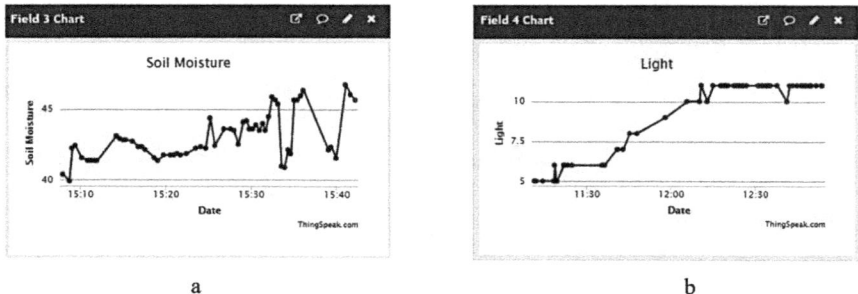

a   b

FIGURE 7.15   (a) Real-time soil moisture (b) Real-time light intensity variation.

Figure 7.15a represents the real-time data soil moisture sent by NodeMCU. Figure 7.15b represents the real-time data of the light intensity of the present day and night. LDR sensor gives output in the range from 0 to 1023. So, the values are mapped from 0 to 99 using map function in Arduino IDE.

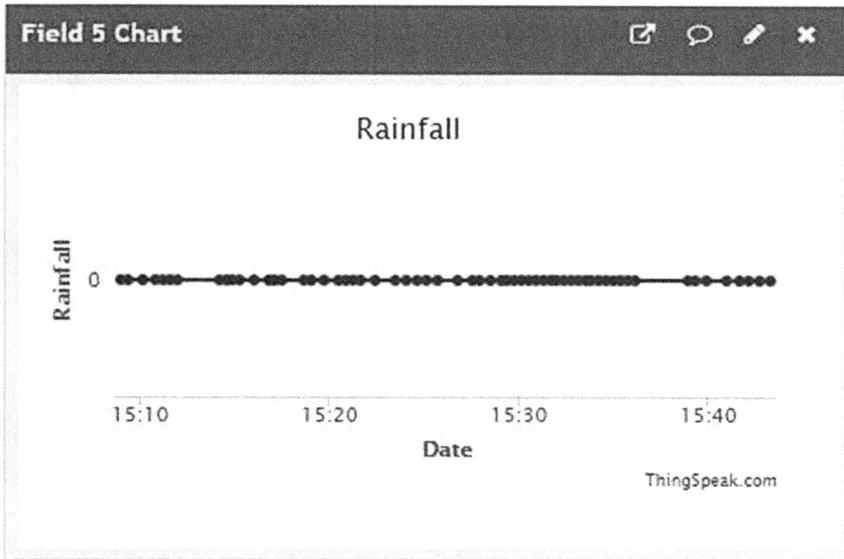

FIGURE 7.16   Real-time rainfall variations.

Figure 7.16 represents the real-time data of the rainfall sensor sent by NodeMCU. If there is any rainfall, then it will send "1" otherwise "0."

7.7.4   PROGRAM CODE FOR NODEMCU

```
#include <DHT.h>
#include <ESP8266WiFi.h>
#define DHTPIN 0 // DHT11 Connection pin
const int sensor_pin = A0; // Analog pin for sensor
#define LDRpin A0 //LDR is connected to D4
String apiKey = "X X X X X X X X X X X X X "; // Put Thingspeak Write
API key here
const char* ssid = "X X X X";   // Enter your WiFi Network's SSID
```

```
const char* pass = "X X X X X X X X"; // Enter your WiFi Network's
Password
const char* server = "api.thingspeak.com";
DHT dht (DHTPIN, DHT11);
WiFiClient cli;
void setup() {
{
Serial.begin ( 115200 );
delay (10);
dht.begin ();}
Serial.println ("Connecting to ");
Serial.println (ssid);
WiFi.begin (ssid, pass);
while (WiFi.status () != WL_CONNECTED)
{   delay (100);
Serial.print ( " " );   }
Serial.println ( " " );
Serial.println ("***WiFi connected***");}
void loop ( ) {
float humi = dht.readHumidity ( );  //DHT 11 Sensor
float temp = dht.readTemperature ( );
float sensorValue = analogRead ( LDRpin );   //Light Sensor
float sensorValue1 = map (sensorValue, 0, 1024 ,30 , 0);
float sensor_analog = digitalRead (sensor_pin);  // Sensor Value read
float moisture_percentage = ( 100 - ( (sensor_analog/1023.00) * 100 ) );
int sensorValue3 = digitalRead ( D1 );  //Rainfall Sensor
float rainSenseReading = ( digitalRead (sensorValue3 ) );
int sensorValue4 = digitalRead ( D4 );  //pH sensor
float pHReading = ( digitalRead ( sensorValue4 ) );
if ( cli.connect (server,80) )  //  "184.106.153.149"
{String sendData = apiKey+"&field1="+String(temp)+"&field2="+String
(humi)+"&field3="+String(moisture_percentage)+"&field4="+String(se
nsorValue1)+"&field5="+String( rainSenseReading )+"&field6="+String
( pHReading )+"\r\n\r\n\r\n\r\n\r\n\r\n";
cli.print ("POST /update HTTP/1.1\n");
cli.print ("Host: api.thingspeak.com\n");
cli.print ("Connection: close\n");
cli.print ("X-THINGSPEAKAPIKEY: "+apiKey+"\n");
```

```
cli.print ("Content-Type: application/x-www-form-urlencoded\n");
cli.print ("Content-Length: ");
cli.print (sendData.length ( ));
cli.print ("\n\n");
cli.print (sendData);
Serial.println ("%. Connecting to Thingspeak."); }
cli.stop ( );
Serial.println ("Sending….");
delay (10000);
}
```

7.7.5   STEPS TO PROGRAM NODEMCU USING DIFFERENT SENSORS AND SENDING THE DATA TO THE CLOUD

Step 1: Import required libraries like #include<DHT11.h> for DHT11 sensor and ESP8266WIFI.h for NodeMCU.

Step 2: Define pins for the sensors either digital or analog.

Step 3: Enter your Wi-Fi Network's SSID & Password.

Step 4: Set the baud rate and check the connection status of whether the NodeMCU is connected to a network or not.

Step 5: Read the values from the sensors using digital Read function and send the data to the cloud, that is, ThingSpeak platform by providing a unique channel id.

Step 6: Set the required amount of delay that how often the NodeMCU should send the data.

7.7.6   PROGRAM CODE FOR PYTHON (MACHINE LEARNING MODELS) FOR K-NN

```
import pandas as pd
import numpy as np
import matplotlib.pyplot as plt
dataset = pd.read_csv('Flower power 6A22.csv')
X = dataset.iloc[:, [1,2]].values
y = dataset.iloc[:, 3].values
```

```
#Preprocessing
nans = data.shape[0] - data.dropna().shape[0]
print("Missing elements are : ", nans)
data.isnull().sum()
cat = data.select_dtypes(include=['O'])
cat.apply(pd.Series.nunique) #It will find unique elements in the dataset
data.Survived.value_counts(sort=True)
data.Survived.fillna('1.0',inplace=True)
from sklearn.model_selection import train_test_split
X_train, X_test, y_train, y_test = train_test_split(X, y, test_size = 0.25,
random_state = 0)
#Feature Scaling
from sklearn.preprocessing import StandardScaler
sc_X = StandardScaler()
X_train = sc_X.fit_transform(X_train)
X_test = sc_X.transform(X_test)
#Clf
from sklearn.neighbors import KNeighborsClassifier
classifier   =   KNeighborsClassifier(algorithm='auto',   leaf_size=25,n_
jobs=1, n_neighbors=25, p=7, weights='uniform', metric = 'minkowski')
classifier.fit(X_train, y_train)
y_pred = classifier.predict(X_test)
#Confusion_matrix
from sklearn.metrics import confusion_matrix
cm = confusion_matrix(y_test, y_pred)
#Applying K-Fold Cross Valid
from sklearn.model_selection import cross_val_score
accuracy = cross_val_score(estimator = classifier, X = X_train, y = y_train,
cv =10)
accuracy.mean()
from sklearn.metrics import roc_auc_score
from sklearn.metrics import precision_score
from sklearn.metrics import recall_score
print(roc_auc_score(y_pred, y_test))
print(recall_score(y_pred, y_test))
print(precision_score(y_pred, y_test))
```

7.7.7 STEPS TO BUILD A MACHINE LEARNING MODEL

Step 1: Import required libraries like NumPy, pandas, and matplot. Numpy is a library used to calculate the mean, median, and other mathematical equations. Pandas is a data frame used to perform an action to display the CSV files in the same format as in the desktop. Matplot is used to plot graphs, charts, bar diagrams, etc.

Step 2: Import the dataset using pd.read_csv function.

Step 3: Set the target column to be classified.

Step 4: Preprocess the data before training the model. Preprocessing involves finding, removing, replacing, sorting of the dataset.

Step 5: Split the dataset into the train and test set. The training set contains nearly 70–80% of the dataset where the testing set contains only 20–30% of the data. The training data are well known to the ML model. The test data are used to check whether the model is well trained or not.

Step 6: Feature Scaling is to be done to maintain the same format or the same weightage of the data to prevent the algorithm from considering greater values and neglecting the smaller ones by giving less priority to the unit.

Step 7: Required classifier like k-NN as shown in the above code is to be imported. Then fit the model into the classifier and predict the result.

Step 8: Confusion matrix is to be used to analyze the performance of the data. The matrix contains TP (True Positive), TN (True Negative), FP (False Positive), FN (False Negative). These values indicate how well the dataset can be used by the user.

Step 9: K-Fold cross-validation is used to check the data within the training set by diving 5% of the training set and making predictions on it.

Step 10: Roc_auc_score, precision_score, recall_score are used to validate the result of the model.

7.7.8 FOR SVM

```
from sklearn.svm import SVC
model = SVC (kernel='linear')
model.fit (X_train,y_train)
```

```
sklearn.svm.SVC  (C=1.0,  kernel='linear',  degree=3,  gamma='scale',
coef0=0.0, shrinking=True, probability=False,tol=0.001, cache_size=200,
class_weight=None, verbose=False, max_iter=-1, random_state=None)
pred = model.predict(X_test)
#K-Fold Cross
from sklearn.model_selection import cross_val_score
accuracy = cross_val_score(estimator = classifier, X = X_train, y = y_train,
cv =10)
accuracy.mean()
from sklearn.metrics import roc_auc_score
from sklearn.metrics import precision_score
from sklearn.metrics import recall_score
print(roc_auc_score(y_pred, y_test))
print(recall_score(y_pred, y_test))
print(precision_score(y_pred, y_test))
```

7.7.9   FOR RANDOM FOREST

```
from sklearn.ensemble import RandomForestClassifier
clf = RandomForestClassifier(n_estimators = 100, criterion = 'gini',
random_state=0, n_jobs = -1)
clf.fit(X_train,y_train)
pred = clf.predict(X_test)
from sklearn.model_selection import cross_val_score   #K-Fold Cross
accuracy = cross_val_score (estimator = classifier, X = X_train, y =
y_train, cv =10)
accuracy.mean()
from sklearn.metrics import recall_score
print (roc_auc_score(y_pred, y_test))
print (recall_score(y_pred, y_test))
print (precision_score (y_pred, y_test))
```

7.7.10   STEPS TO BUILD THE ABOVE ML MODELS

Step 1: Steps 1–6, Steps 8–10 from 6.3.2 is to be followed for all the ML models to import, preprocess, and for feature scaling.

Step 2: Necessary libraries for the model are to be imported from the sklearn library.

Step 3: Steps 8–10 from 6.3.2 is the same for all the ML models.

7.8 RESULT AND ANALYSIS

From this project, we can able to understand that how IoT and machine learning provide some sort of action and help in the agricultural sector. Three algorithms like k-NN, SVM, and Random Forest are used in this project. The expected accuracy for Random Forest is 85, for SVM 84 and k-NN it will be 88%.

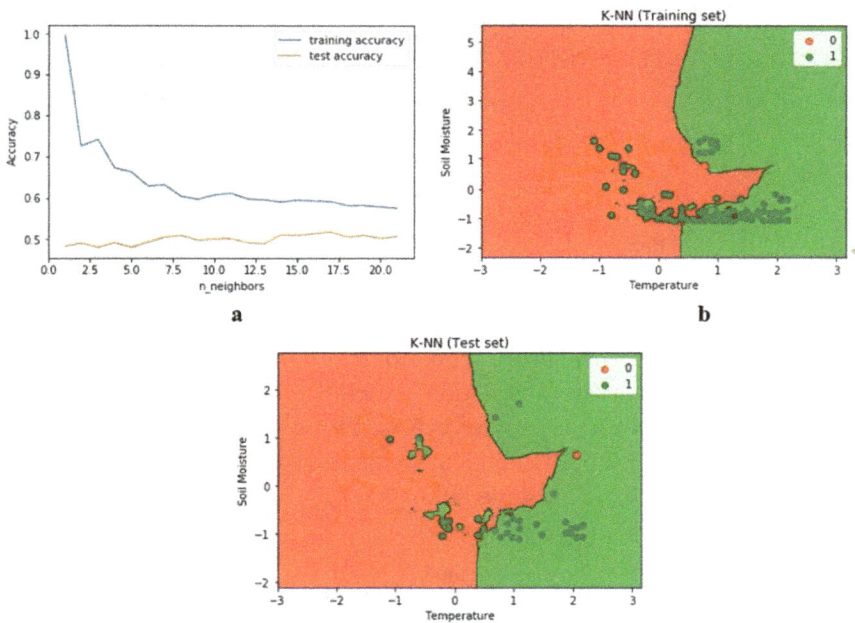

FIGURE 7.17 (a) Train v/s test accuracy (b) Prediction on training set for (k-NN) (c) Prediction on test set for (k-NN).

In Figure 7.17a, the graph represents the train and test accuracy for the dataset. It shows that the unbiased situation provides good results by avoiding overfitting or leakage of data under any circumstances. In Figure

7.17b, the number of 0's and 1's are plotted based on two parameters, that is, Temp and Soil Moisture. Green color dots are 1's and Red color dots are 0's. The line that divides the green and red dots is known as prediction line. One part contains the dots of other parts and the accuracy depends on how the prediction line divides the points exactly. In Figure 7.17c, the prediction line will divide the two categories mostly in two equal parts. This determines that the model is running fine and the prediction results are accurate (Fig. 7.18).

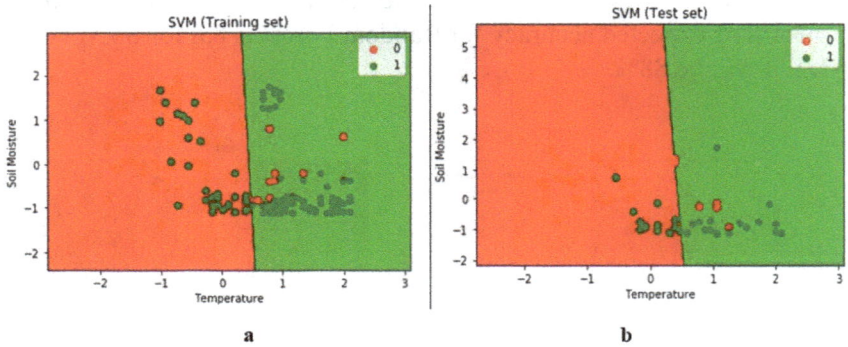

FIGURE 7.18   (a) Prediction on training set for SVM (b) Prediction on test set for SVM.

In the above two figures, the line that divides the two categories is known as hyperplane which was commonly seen in SVM. This is a kernel SVM in which Radial Basis Function is used as a kernel. The line is drawn with the help of various support vectors. These support vectors in SVM will calculate the distance (Euclidean distance) with the other neighboring vectors and verify which sample belongs to which category (Table 7.2).

TABLE 7.2   Result of the Proposed Methods.

S. no	Machine learning models	k-NN (%)	SVM (%)	Random forest (%)
1	Training accuracy	89	85	86
2	Testing accuracy	87	84	83
3	Prediction accuracy	88	84	85
4	Error (%)	12	16	15
5	Area under curve	90	87	89

TABLE 7.2   *(Continued)*

S. no	Machine learning models	k-NN (%)	SVM (%)	Random forest (%)
6	Specificity = Tn/(Tn + Fp)	88	83	87
7	Precision (P) = Tp/(Tp + Fp)	92	88	84
8	Sensitivity (or) Recall [R] = Tp/(Tp + Fn)	89	85	86
9	F1 score 2(P.R)/(P + R)	90	87	85

The above table compares the results obtained between k-NN, SVM, and Random Forest. It shows that the prediction accuracy is more for k-NN. The area under Curve (AUC/ROC curve) is a parameter that tells us how much data are useful. The Error tells that how much data are unused by the model. Specificity, Precision and F1 score tell us that how our model is being performed. If those values are good then our model performance is good. Specificity avoids false alarms and false positives. Precision is to be chosen if our measurement meets the true value. If FP are better than FN then Recall will be chosen to analyze the model. Out of all the three model performances, k-NN provides the best results in terms of fewer error results and accuracy. The results can be improved by using different methods in the models like parameter tuning or by building our custom models which give better accuracy.

7.9   SUMMARY AND CONCLUSIONS

The proposed architecture was implemented and evaluated. By this, we can conclude that this chapter describes the architecture of how IoT and machine learning play an impactful role in the area of agriculture. The application of this system in this area will advance the harvest and production of the crop. By implementing IoT in agriculture, there will be less useful in manpower and the cost of labor is also reduced. The whole system is efficient and consumes less power. The evaluation of each model is identified and compared with other models. We vary the size of the datasets that depend on the type of soil for better accuracy. The highest accuracy is found to be 90% in a Loamy type of soil. By this, k-NN shows the best accurate result comparison to all the other two classifiers. In the future, weed detection technique using ESP -32 cam, and fertilizer prediction required for the soil is my concern. By adding LoRa (long-range) technology to this system will help the user to get the results that are far away from some kilometers.

KEYWORDS

- IoT
- machine learning
- SVM
- KNN
- random forest
- node MCU

REFERENCES

1. Zhao, J-C. et al. The Study and Application of the IoT Technology in Agriculture. *2010 3rd International Conference on Computer Science and Information Technology*, Vol. 2; IEEE, 2010.
2. Patil, K. A.; Kale, N. R. A Model for Smart Agriculture Using IoT. *2016 International Conference on Global Trends in Signal Processing, Information Computing and Communication (ICGTSPICC)*; IEEE, 2016.
3. Pallavi, S.; Mallapur, J. D.; Bendigeri, K. Y. Remote Sensing and Controlling of Greenhouse Agriculture Parameters Based on IoT. *2017 International Conference on Big Data, IoT and Data Science (BID)*; IEEE, 2017.
4. Bing, F. Research on the Agriculture Intelligent System Based on IoT. *2012 International Conference on Image Analysis and Signal Processing*; 2012.
5. Davcev, D. et al. IoT Agriculture System Based on LoRaWAN. *2018 14th IEEE International Workshop on Factory Communication Systems (WFCS)*; IEEE, 2018.
6. Math, R. K. M.; Dharwadkar, N. V. IoT Based Low-cost Weather Station and Monitoring System for Precision Agriculture in India. *2018 2nd International Conference on I-SMAC (IoT in Social, Mobile, Analytics and Cloud)(I-SMAC) I-SMAC (IoT in Social, Mobile, Analytics and Cloud)(I-SMAC), 2018 2nd International Conference on*; IEEE, 2018.
7. Rahman, Sk Al Z.; Mitra, K. C.; Islam, SM. M. Soil Classification Using Machine Learning Methods and Crop Suggestion Based on Soil Series. *2018 21st International Conference of Computer and Information Technology (ICCIT)*; IEEE, 2018.
8. Chlingaryan, A.; Sukkarieh, S.; Whelan, B. Machine Learning Approaches for Crop Yield Prediction and Nitrogen Status Estimation in Precision Agriculture: A Review. *Comput. Electr. Agric.* **2018,** *151,* 61–69.

9. Prakash, S.; Sharma, A.; Sahu, S. S. Soil Moisture Prediction Using Machine Learning. *2018 Second International Conference on Inventive Communication and Computational Technologies (ICICCT)*; IEEE, 2018.

10. Goldstein, A. et al. Applying Machine Learning on Sensor Data for Irrigation Recommendations: Revealing the Agronomist's Tacit Knowledge. *Precision Agric.* **2018,** *19* (3), 421–444.

11. Charoen-Ung, P.; Mittrapiyanuruk, P. Sugarcane Yield Grade Prediction using Random Forest and Gradient Boosting Tree Techniques. *2018 15th International Joint Conference on Computer Science and Software Engineering (JCSSE)*; IEEE, 2018.

12. Medar, R. A.; Rajpurohit, V. S.; Ambekar, A. M. Sugarcane Crop Yield Forecasting Model Using Supervised Machine Learning. *Int. J. Intell. Syst. App.* **2019,** 11 (8), 11.

13. Dholu, M.; Ghodinde, K. A. Internet of Things (IoT) for Precision Agriculture Application. *2018 2nd International Conference on Trends in Electronics and Informatics (ICOEI)*; IEEE, 2018.

14. Dagar, R.; Som, S.; Khatri, S. K. Smart Farming–IoT in Agriculture. *2018 International Conference on Inventive Research in Computing Applications (ICIRCA)*; IEEE, 2018.

15. Heble, Soumil, et al. A Low Power IoT Network for Smart Agriculture. *2018 IEEE 4th World Forum on Internet of Things (WF-IoT)*; IEEE, 2018.

16. Hu, X.; and Qian, S. IoT application system with crop growth models in facility agriculture. *2011 6th International Conference on Computer Sciences and Convergence Information Technology (ICCIT)*; IEEE, 2011.

17. Sushanth, G.; Sujatha, S. IOT Based Smart Agriculture System. *2018 International Conference on Wireless Communications, Signal Processing and Networking (WiSPNET)*; IEEE, 2018.

18. Priya, P.; Muthaiah, U.; Balamurugan, M. Predicting Yield of the Crop Using Machine Learning Algorithm. *Int. J. Eng. Sci. Res. Technol.* **2018,** *7* (1), 1–7.

19. Dhall, R.; Agrawal, H. An Improved Energy Efficient Duty Cycling Algorithm for IoT Based Precision Agriculture. *Procedia Comput. Sci.* **2018,** *141*, 135–142.

20. Vanitha, C. N.; Archana, N.; Sowmiya, R. Agriculture Analysis Using Data Mining and Machine Learning Techniques. *2019 5th International Conference on Advanced Computing & Communication Systems (ICACCS)*; IEEE, 2019.

21. Jaiganesh, S.; Gunaseelan, K.; Ellappan, V. IoT Agriculture to Improve Food and Farming Technology. *2017 Conference on Emerging Devices and Smart Systems (ICEDSS)*; IEEE, 2017.

22. Prathibha, S. R.; Hongal, A.; Jyothi, M. P. Iot Based Monitoring System in Smart Agriculture. *2017 International Conference on Recent Advances in Electronics and Communication Technology (ICRAECT)*; IEEE, 2017.

23. Reshma, S. R. J.; Pillai, A. S. Impact of Machine Learning and Internet of Things in Agriculture: State of the Art. *International Conference on Soft Computing and Pattern Recognition*; Springer, Cham, 2016.

24. Mekala, M. S.; Viswanathan, P. A Novel Technology for Smart Agriculture Based on IoT with Cloud Computing. *2017 International Conference on I-SMAC (IoT in Social, Mobile, Analytics and Cloud) (I-SMAC)*; IEEE, 2017.

25. Khanna, A.; Kaur, S. Evolution of Internet of Things (IoT) and Its Significant Impact in the Field of Precision Agriculture. *Comput. Electr. Agric.* **2019,** *157*, 218–231.

26. Oliveira, I. et al. A Scalable Machine Learning System for Pre-Season Agriculture Yield Forecast. arXiv preprint arXiv:1806.09244, 2018.

27. Treboux, J.; Genoud, D. Improved Machine Learning Methodology for High Precision Agriculture. *2018 Global Internet of Things Summit (GIoTS)*; IEEE, 2018.

28. Shinde, S.; Kulkarni, M. Review Paper on Prediction of Crop Disease Using IoT and Machine Learning. *2017 International Conference on Transforming Engineering Education (ICTEE)*; IEEE, 2017.

29. Karim, F.; and Karim, F. Monitoring System Using Web of Things in Precision Agriculture. *Procedia Computer Science* **2017,** *110*, 402–409.

30. Ghadge, R. et al. Prediction of Crop Yield Using Machine Learning. *Int. Res. J. Eng. Technol. (IRJET)* **2018,** *5*.

CHAPTER 8

Bluetooth Robotics

P. RAJA[*1], DUSHYANT KUMAR SINGH[2], and HIMANI JERATH[3]

School of Electronics & Electrical Engineering, Lovely Professional University, Phagwara, Punjab, India

Corresponding author. E-mail: rajsella.visys@gmail.com

ABSTRACT

The onset of the 21st century marked the automation and remote controlling of most commercial products. Human beings are expecting help from manmade machinery/robots to complete the specific task without any human intervention. So, it is very important that to design such kinds of modules/robots which will perform predefined tasks or are controlled by remote to perform the particular task like in the field of healthcare, education, industry, transport, agriculture, retail, and etc. This chapter discusses the Bluetooth-controlled robot that will receive the command from the Arduino board with help of a Bluetooth module and Bluetooth terminal application which is installed in our smartphones. The control signals/commands are designed by the programers in such a way that they can provide the guidance/control signal to the robot.

8.1 INTRODUCTION

Recently, there was a lot of risk on the workers in the field of fire extinguishing. The fire-fighting robots can be used to protect fire-extinguishing personnel of the risk of combustion and inhalation of toxic gases and explosive materials, especially in confined areas and narrow. This robot is leading to the maintenance of life of workers in the field of fire extinguishing. A Bluetooth control robot system is an electromechanical device used in industry to replace human work to carry out the functions assigned to him. It can interact with its environment; sometimes it may resemble a human being physically or carry out its tasks in a human way.[5]

Versatile Robots: is a framework equipped for moving their bodies from one spot to another in its condition. Portable robots come in two assortments: fastened and self-sufficient. A fastened robot by dumping its capacity supply and cerebrum over the edge, conceivably depending on a work station and a divider outlet. Control signals and force are gone through a heap of wires (the tie) to the robot, which is allowed to move around in any event to the extent the tie will permit.[5]

Self-governing portable robots: These assortments of robots are needed to carry everything alongside them, including a force supply and a cerebrum. The force supply is typically numerous batteries, which adds a ton of weight to the robot. The mind is likewise limited since it must fit on the robot, and be frugal about ingestion power out of the batteries.[5]

Today, Robotics has performed most prominent accomplishment in the realm of mechanical assembling. Robot legs or mechanical hands are progressively adaptable and can move at a particular situation in the sequential construction system, the robot leg can move quickly with precision to perform redundant assignments, for example, painting and welding.

In any case, with every one of these victories, the business robots experience the ill effects of a basic disservice: absence of versatility. A fixed leg has a constrained scope of development, which relies upon where it pulled down. In other hand, a versatile robot would have the option to development all through the assembling plant, applying deftly their gifts any place it is best.[5]

A mobile robot is a system, which has following characteristics:

• Total mobility is relatively to the environment.
• Human interaction is less.
• Sense and react to the environment variable.

Today, we all are in the kingdom of mechanical autonomy. Purposely or unwittingly, we had been utilizing various sorts of robots in day-by-day life. The point of the philosophy is to assess what understudies can find out about the arenas of building, mechatronics, and programing advancement as they configure, develop, and program an independent robot. This will give a rule to the understudies who are new in the realm of Arduino and help them to comprehend about implanted framework, microcontroller object detector sensors are used to develop a robot using Arduino.[1]

This chapter will result a robot that can move with the instructions from your android phone and can go to a remote abstruse place to collect data and help people. The Robot will be Bluetooth-controlled having 360 degree movement and can be controlled from a distance place remotely. The robot can able to get the instructions from the mobile throw the Bluetooth module HC-05. Then it will process the instruction through microcontroller Arduino and move the robot by following the instructions. Then it will collect data from there and send the collected data to cloud database via Wi-Fi and internet.

In my system, I have designed a low-cost Microcontroller-Based Android controlled Robot. The robot will move forward, backward, left and right direction by following the instructions given from the mobile. This system can be helpful for various purposes.[1]

As of now, the improvement of self-sufficient robots has been getting a lot of consideration from analysts around the world. In any case, not all tenders need the robot to be entirely mechanized. Some specific automated requests want human arbitration for instance looking over excavation site. The robots have been made this much advance that they can adapt to the client's need easily. Like robots can yield a particular area, estimates the parameters like temperature, moistness, and other comparative assignments. In other conditions, for example, confronting impediment while moving, the robot ought to have the option to evade the deterrent itself before accepting consequent control activities from client. Accordingly, it is critical to create semi-mechanized versatile controlled robot with obstruction shirking ability.[2]

Bluetooth-controlled robots are programable remotely monitored/ controlled machines that are intended to perform at least one assignment consequently with speed and exactness. Because of its ability to work persistently, creation will be expanded. Cell phones have become ground-breaking gadgets with quicker processors, more brilliant devices, and more specialized techniques. Nowadays, signals could be utilized to interface with machineries. A signal is a type of nonverbal correspondence where developments of body parts convey specific messages. Current cell phones are implanted with the accelerometer sensor, Bluetooth segment and are controlled by Android. Bluetooth controlled Robots could be controlled by control signals.[3]

Presently a-days, apply autonomy is getting one of the most developed in the field of innovation. The uses of development characterize and furthermore utilized as a putting out fires robot to help the individuals from the fine mishap. Be that as it may, governing the robot with an isolated of adjustment is very confounded. Along these lines, another task is built up that is an accelerometer-based signal control robot. The fundamental objective of this undertaking is to control the development of the robot with a hand utilizing Bluetooth. Right now, the android advanced cell is utilized as a remote control for working the robot.[4]

Robots are replacing human assignments and occupations. They will work in human living condition and not simply in a fixed work environment for the most part with robots. This brings robot security as one of the most significant research areas while building the agriculture-based robot application. A contextual analysis to show a more secure robot by applying an ultrasonic sensor at the front of the robot to forestall crash while progressing at high speed. Bluetooth is a typical remote innovation in a large portion of cell phones today; so, we fabricate a fundamental two-wheeled robot constrained by an advanced mobile phone by means of Bluetooth correspondence. Contrast with infrared control module, Bluetooth is fit for point to multifocus associations and does not require view. There will be more prospects when we require extra highlights as in on the present robot control framework. Another bit of leeway is that Bluetooth can trade information so the natural information recognized by the robot can be send to another Bluetooth gadget for information assortment. This information can likewise help clients to give suitable orders when the robot is far out. As in driving, we ought to keep up a protected after separation which changes at various speeds. The sheltered separation used to trigger the counter crash framework is determined dependent on the present speed of the robot. The progress ahead is additionally bolted up when the front separation needs more space to maintain a strategic distance from mishap by human control blunder. All the occurrences experienced are sent back to the advanced mobile phone for helping the end client to work the robot appropriately without impact. Ultrasonic sensor is applied to the framework like, however, we just require one sensor which is adequate for the general element of our robot.[7]

With the extension of the web, the information world is extending. In any case, person cannot live on information alone with genuine sense. Taking along the universe and the physical world is continuously expanding and there is a need to grow along with it for the upgraded information step by step. Unique of the outcomes is Internet empowered automaton that can be precised over the web by an edge among the Bluetooth-controlled robot and microcontroller.

Primary of its sort utilized Ethernet links to govern the robot. Be that as May, it was badly designed as to governor robot with cable and relevant to restricted separation as it were. So steadily, it moved to remote LAN control. Be that as it may, the issue with remote LAN control is its constrained ability to deal with information and variety. To evacuate of this weakness utilization of cell GSM part to speak with the robot remained existing. It remained helpful as a result of the remote framework empowers to work in go as long as there was a cell organization.

In any case, the cell module cannot move the large informational collection momentarily. Subsequently, controlling the robot and taking care of information all the while was impractical easily that presented huge requirements in utilizing this framework. Meanwhile, 3G versatile innovation was presented that had a major effect in controlling robot contrasted with its antecedents. With 3G, a huge measure of information can be moved in brief timeframe. In this way, it was conceivable to control and get criticism all the while. In the present day, Android and Arduino are utilized; however, the issues stay in these frameworks as these robots require complex equipment and some of the time even their own servers. In addition, proficient aptitudes are important to control these robots which are regularly rare toward the end client level. Moreover, these are not cost-productive, uncommonly, in a creating nation like Bangladesh where robots are being utilized in a constrained scale.[10]

In today's world, people are concerning about effective, automatic, and economic. The competitive environment makes people busy with their working and to a certain extent, floor cleaning in their house considered time-wasting. However, not everyone is affordable to buy an automatic floor cleaner due to its price. The second factor that makes the automatic floor polisher and cleaner less popular in the market is the size. Some of them are big in size to have all unnecessary functions on it.

In fact, most of us are usually using sweeper for cleaning. From time to time, technologies come up and need to upgrade to ease human task. In addition, most of the people are busy working and they do not have enough time to clean their houses which in turn make an automatic floor cleaner very attractive in the market.[14]

8.2   HARDWARE IMPLEMENTATION

In our real-world, nowadays people are becoming very lazy and a lot of ignorance toward to do their routine work. In order to solve the issue, we need support from manmade system/modules which will help or assist to people in various fields. The systems/modules are designed in this it can sense and do some process on the detected values, if needed the system will reproduce the output signal in necessity format. In this modern world, we have a lot of latest/updated components, modules, advanced controllers, and intelligent sensor are available with help of that we can develop and design automated and intelligent systems. In this chapter, we discuss about Bluetooth-controlled robotics which can be used for different applications like Low variety Mobile Investigation Plans Military Bids (no human intervention) Assistive campaigns (like wheelchairs) Home mechanization.

The robot can serve the society. They can collect data from remote place and patients. Evaluated wellbeing will be eventual fate of medicinal services since wellbeing that is quantifiable can be better improved. In this way, it is shrewd to exploit measured wellbeing innovation. We likewise realize that information influences execution in this way, an item estimation and following of wellbeing for better results is required. Health is considered as one of the basic human needs. If sound health of mass people can be guaranteed, then the overall productivity is obvious to enhance. This will have a great positive impact on our society.[1]

Today, we are in the realm of mechanical technology. Intentionally or unwittingly, we have been utilizing various kinds of robots in our everyday life. The point of the theory is to assess what understudies can find out about the fields of building, mechatronics, and programing improvement as they configure, develop, and program an independent robot. This will give a rule to the understudies who are new in the realm of Arduino and help them to comprehend about installed framework, IR sensors, microcontroller, and how to make a robot utilizing Arduino.

There is a complete straightforward framework of the structure. Parameters like cost viability, low profile structure, and effortless in configuration have been taken into account in the preplanning phase. Our framework intends to accomplish such objective to not only plan a framework that can give variety of functionalities, but also a straightforward and simple to-utilize interface that can be user friendly as well.

Here, an android application will be built that will be acting as a Portable Remote Control. The main emphasis is on the trending innovation as a combination of both mobile and robot commonly known as "Mobot." A mobile phone on an android platform can be considered along with its innovation and to collaborate with the inserted framework. In near future, these multifunctional versatile robots and Bluetooth are the on-going advances that can be utilized to support mankind. This system consists of Arduino Uno—a microcontroller based prototyping board, classic Bluetooth communication module, and DC motor driver chip. Bluetooth communication module HC-05 is interfaced with Arduino UBO main board is to be utilized for the client's application. The system comprises of giving the commands from mobile phone-based Bluetooth application to Arduino Uno. The command so received is processed by the Arduino as per the application program into it. After processing the command so received, Arduino generated the control signal which goes to the driver IC, which in turn obtains the control of the specific DC motor. The main components of the present system comprise of Arduino, an open source prototyping board, a classic Bluetooth communicating module, that is, HC-05, motor driver IC to provide sufficient excitation of motor to rotate and smart phone with RC controller mobile app installed. The elementary construction lumps of the project have been labeled below:

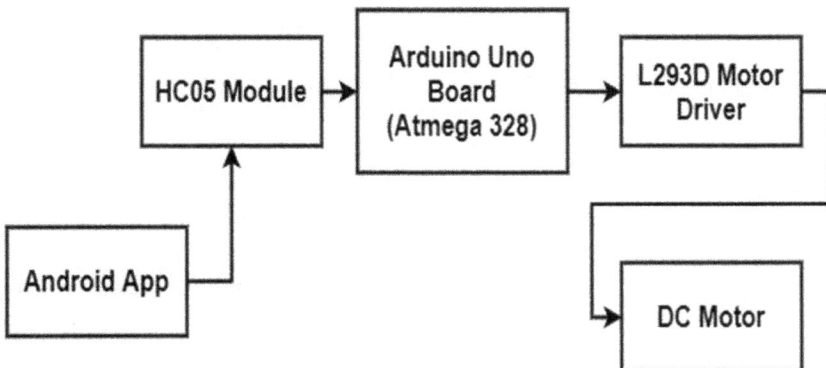

FIGURE 8.1   General block diagram.

Microcontroller becomes the heart and mind of the robot. Presently, microcontroller-based board Arduino Uno is being used that has ATMEGA

328P as a main processing chip as shown in Figure 8.1. The application for microcontroller used in Arduino is coded using embedded C-programing language. The programing in Arduino is stored in its Read Only Memory (ROM). The smart cell phone is used and the transmitter and is used to send the command information to microcontroller through Bluetooth module. Basically, smart phone with installed RC controller application, as given in Figure 8.2, acts as distant control unit, that is, remote for the present RC-controlled robot application. The smart phone sends the guidance to the robot in the form of predefined commands such as forward, reverse, left, and right. The upside of the present work is the android programing which has been kept simple but including all the essential capabilities. The oddity lies in the effortlessness of the plan and working.

DC motor is an electric device which is capable for converting electric energy into mechanical movement. DC motor is based on the concept that when a current carrying conductor in the magnetic field, torque is generated. In DC motor by shifting the current form clockwise to anti—clockwise, the direction of rotation of motor can be changed. Speed of motor is controlled by controlling the amount of current flowing to DC motor.

Bluetooth is developed by Sweden Company Eriksson. Bluetooth works in the Industrial, Scientific and Medical (ISM) standard in short range recurrence band of 2.4 GHz with each channel particularly in 2400–2483.5 MHz recurrence band, which incorporates watch groups too. It uses the technology of FHSS (Frequency Hop Spread Spectrum) where information parcels are partitioned dependent on recurrence more than 79 assigned Bluetooth channels. Each channel has a transfer speed of 1 MHz. The fresher Bluetooth 4.0 standard, notwithstanding, utilizes 2 MHz steps and along these lines has 40 assigned channels. It utilizes a variety of FHSS called Adaptive Frequency-jumping spread range (AFH), which hypothetically skips channels with obstruction and results in better correspondence.

Bluetooth is basically a convention with an ace slave design, which implies that one ace gadget can speak with up to seven gadgets. This was and is a gigantic bit of leeway to before wired correspondence conventions which could work just with a one to one arrangement. Basically, making another standard called Personal Area Networks (PANs), Bluetooth achieved unmistakably progressively successful impromptu systems and permits correspondence without conventional host-based systems administration. This system of Bluetooth gadgets is known as a "piconet."

There's likewise work proceeding to make something many refer to as a "disperse net," which is a mix of at least two piconets, where a gadget that goes about as an ace in one piconet can be a slave in another.

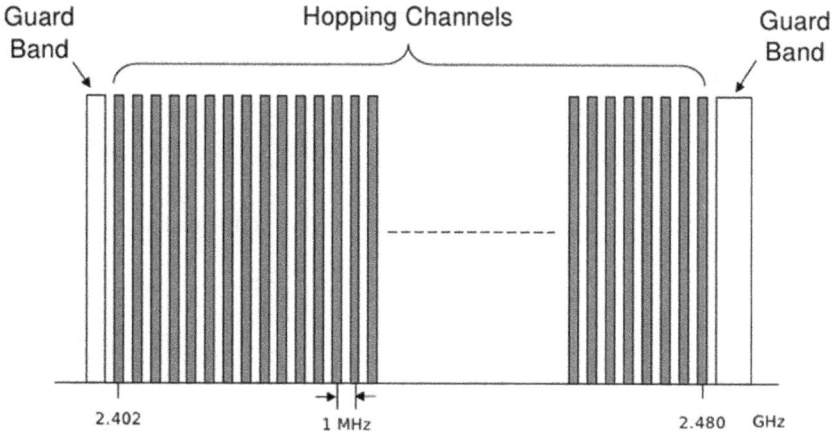

FIGURE 8.2 Band diagram.

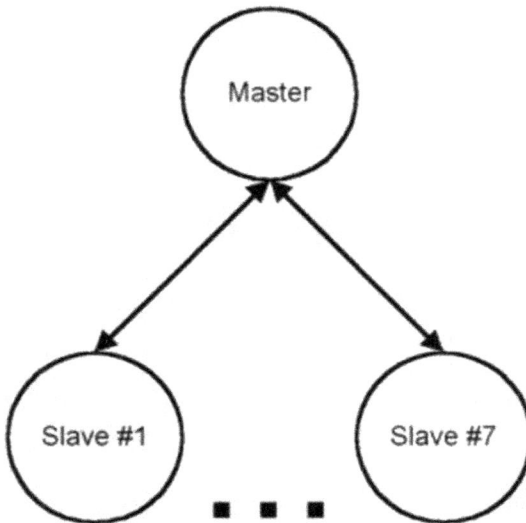

FIGURE 8.3 Master slave configuration.

Most Bluetooth gadgets have a scope of around 10 m or something like that since they're battery worked and are of the Class 2 sort. Contingent upon power utilization and range, there are fundamentally three kinds of Bluetooth gadgets. They are:

Classic Bluetooth is mainly divided into three classes based on power and their communication range. Most of the classic Bluetooth devices have reach up to 10 m as they all are of class 2. These three Bluetooth classes are:

- Class 1 with power 100 mW and communication distance range of about 100 m
- Class 2 with Power 2.5 mW and communication distance range of about 10 m
- Class 3 with Power 1 mW and communication distance range of about 1 m

The drop in the range form 10 m to only 10 m from class 1 is due to the gigantic drop in the transmission power but for day-to-day use 10 m of range is sufficient. For bigger applications and impressive handset range of 100 m can be obtained and these devices substitute even Wi-Fi in certain circumstances such as applications where WLAN is required.

With the time, Bluetooth technology has gone from ambiguous system convention to one of the most notable and ordinarily utilized models on the planet. Below are the different development phases for Bluetooth technology:

- Bluetooth version 1.0—The principal standard presented, it was not really utilized monetarily in light of the fact that it experienced issues with interoperability, which should be the primary draw of a widespread correspondence standard.
- Bluetooth version 1.1—Dubbed IEEE 802.15.1-2002 since the SIG did not exist, it fixed a great deal of the issues from the past forms and included non-encoded channels and sign quality pointers.
- Bluetooth version 1.2—Brought about a lot of quicker exchange speeds, presented AFH and better transmission shows, for example, retransmission of adulterated information bundles.
- Bluetooth version 2.0 + EDR—Again, achieved quicker exchange speeds, upto 3 Mbits/s hypothetically. EDR meant "Upgraded Data Rate," to connote this.

- Bluetooth version 2.1 + EDR—A significant update that let gadget blending happen a lot quicker and all the more effectively, as we probably are aware today.
- Bluetooth version 3.0 + HS—Another significant amendment that permitted information move of upto 24 Mbits/s, however, not on the real Bluetooth channel. The Bluetooth channel was utilized to match gadgets, at that point the genuine exchange was done over a channelized Wi-Fi connect.
- Bluetooth version 4.0 LE—Called Bluetooth Low Energy, this definitely brought downforce required while keeping information rates up, which opened up a totally different universe of continually associated gadgets, for example, wellness groups, savvy watches, and such. It was absurd before to keep the connection on for a really long time as a result of battery and warmth issues, so Bluetooth v4.0 LE was something of an achievement.
- Bluetooth version 4.1—An advancement of v4.0, this variant will bolster LTE moves, higher trade rates, better security conventions, effective blending, and diminished system cycles.

Bluetooth offers help for three general application territories utilizing short-extend remote availability:

- Data and voice passages—Bluetooth encourages constant voice and information transmissions by giving easy remote association of compact and stationary specialized gadgets.
- Cable substitution—Bluetooth disposes of the requirement for various, regularly exclusive link connections for the association of essentially any sort of specialized gadget. Associations are moments and are kept up in any event, when gadgets are not inside view. The scope of each radio is roughly 10 m, however, can be stretched out to 100 m with a discretionary speaker.
- Ad hoc organizing—A gadget furnished with a Bluetooth radio can build up moment association with another Bluetooth radio when it comes into go.

8.2.1 ASSOCIATION PROCESS

Making a Bluetooth association between two gadgets is a multistep process including three dynamic states: Inquiry—If two Bluetooth gadgets know literally nothing about one another, one must run a request to attempt to find the other. One gadget conveys the request demand, and any gadget tuning in for such a solicitation will react with its location, and perhaps its name and other data.

Paging (Connecting)—Paging is the way toward framing an association between two Bluetooth gadgets. Before this association can be started, every gadget has to know the location of the other (found in the request procedure).

Connection—After a gadget has finished the paging procedure, it enters the association state. While associated, a gadget can either be effectively taking part or it very well may be placed into a low force rest mode.

Active Mode—This is the normal associated mode, where the gadget is effectively transmitting or getting information.

Sniff Mode—This is a force sparing mode, where the gadget is less dynamic. It will rest and just tune in for transmissions at a set interim (e.g., each 100 ms).

Hold Mode—Hold mode is a transitory, power-sparing mode where a gadget rests for a characterized period and afterward returns back to dynamic mode when that interim has passed. The ace can order a slave gadget to hold.

Park Mode—Park is the most profound of rest modes. An ace can order a captive to "park," and that slave will get latent until the ace tells it to wake back up.

8.2.2 HOLDING AND PAIRING

At the point when two Bluetooth gadgets share an extraordinary fondness for one another, they can be fortified together. Fortified gadgets consequently set up an association at whatever point they are sufficiently close. At the point when I fire up my vehicle, for instance, the telephone in my pocket quickly interfaces with the vehicle's Bluetooth framework since they share a bond. No UI connections are required!

Bonds are made through one-time a procedure called matching. At the point when gadgets pair up, they share their addresses, names, and profiles, and for the most part store them in memory. They likewise share a typical mystery key, which permits them to bond at whatever point they are as one later on.

Blending as a rule requires a confirmation procedure where a client must approve the association between gadgets. The progression of the validation procedure changes and as a rule relies upon the interface capacities of one gadget or the other. Here and there blending is a straightforward "Just Works" activity, where the snap of a catch is everything necessary to match (this is basic for gadgets with no UI, similar to headsets). Different occasions blending include coordinating six-digit numeric codes. More seasoned, inheritance (v2.0 and prior), blending forms include the entering of a typical PIN code on every gadget. The PIN code can run long and intricacy from four numbers (e.g., "0000" or "1234") to a 16-character alphanumeric string.

8.2.3   REMOTE COMPARISON

Bluetooth is a long way from the main remote convention out there. You may be perusing this instructional exercise over a Wi-Fi arrangement. Or on the other hand, perhaps you've even played with ZigBees or XBees. Anyway, what makes Bluetooth not the same as the remainder of the remote information transmission conventions out there?

How about we thoroughly analyze. We will incorporate BLE as a different substance from Classic Bluetooth (Table 8.1).

TABLE 8.1   Comparison between Wireless Technologies.

S. no	Particular	Bluetooth	NFC	Infrared	ZigBee	Wi-Fi
1	IEEE standard	802.15.1	802.2	802.11	802.15.4	802.11a/b/g
2	Band of frequency	2.4 Ghz	13.56 Mhz	875 nm	868/915 Mhz, 2.4 Ghz	2.4 Ghz, 5 Mhz
3	Range of coverage	1–100 m	<0.2 m	0.2—1 m	10—100 m	100 m
4	Number of RF channels	79	1	50	1/10, 16	14
5	Data transfer	3 Mbit/s	424 Kbit/s	4 Mbit/s	250 Kbit/s	54 Mbit/s
6	Spread spectrum	FHSS	NA	PPM	DSSS	DSSS, CCK, OFDM

TABLE 8.1   *(Continued)*

S. no	Particular	Bluetooth	NFC	Infrared	ZigBee	Wi-Fi
7	Topologies	Piconet	Point to point	Point to Point	Star	BSS
8	Data protection	16-bit CRC	32-bit CRC	16-bit CRC	16-bit CRC	32-bit CRC

Bluetooth is not the best decision for each remote activity out there, yet it excels at short-go link substitution type applications. It additionally flaunts a regularly more advantageous association process than its rivals (ZigBee explicitly).

ZigBee is regularly a decent decision for checking systems—like home mechanization ventures. These systems may have many remote hubs, which are just meagerly dynamic and never need to send a great deal of information.

BLE joins the comfort of great Bluetooth and includes essentially lower power utilization. Right now, it can become a rival of ZigBee for battery life. BLE cannot contend with ZigBee as far as system size, yet for a single gadget-to-gadget network it's entirely practically identical.

Wi-Fi is likely the most well-known of these four remote conventions. We all are entirely acquainted with what reason for existing it is best for: Internet(!). It is quick and flexible, yet in addition requires a great deal of intensity. For broadband, Internet gets to it destroys different conventions.

The course of revolution will rely upon the extremity of current. This is an engine driver IC that can drive two engines all the while. L293D has 16 pins, and it permits the engines to move in either course. L293D and L293NE are the most ordinarily utilized ICs of this arrangement. Controlling of two DC engines all the while should be possible utilizing these ICs. Essentially, it receives two H-connect which is utilized to control the current of engine.

The engine drivers will be alluded as L293D as it were. With the assistance of these engine drivers, two DC engines can be joined to a solitary IC and then two are moved in two unique ways. It has 16 computerized pins that can be utilized as information and yield. This driver can drive both high and low rpm engines. To choose the course of engine's turn, the voltage might be changed. The structure of robot comprises of Arduino UNO microcontroller, an engine driver L293D, 2 DC engines, a HC-05 (Bluetooth module), and a RFID sensor.

The procedure begins with the exchange of information from Android telephone to HC-05, a Bluetooth module which further exchanges the information to Arduino UNO board. Here, the information is controlled and plays out the capacity which is a controller that controls the signs and plays out the given capacities, and it advances the signs advising the engines which bearing to turn signals. The client can work the robot by utilizing cell phone-based application. Utilizing the applications development as well as different elements of robot will be chosen.

The system has been utilized in various fields like education, industry, military, medical, logistic, and smart Agriculture. Some of various applications are to be described as given below.

8.3   VARIOUS APPLICATION OF BLUETOOTH ROBOTICS

1.   Bluetooth controlled Robot
2.   Bluetooth controlled obstacle avoidance Robot
3.   Military application (No human invention)
4.   Assistive device (like wheelchair)
5.   Home automation

8.4   BLUETOOTH CONTROLLED ROBOT

The chapter discussed the design of smartphone-controlled automated robot over Bluetooth link. Here, the idea is to control a robot through smart phone using Bluetooth as communication link between smartphone and robot. In the robot, the most popular Arduino UNO is used as main processing board and for Bluetooth communication module HC-05 is used.

The remote-control feature of the robot is implemented through android smart phone utilizing the inbuilt Bluetooth facility present in it. Here, the fully equipped inbuilt Bluetooth-based Smartphone is acting as a Remote controlling device for the robot. The main controlling unit of the robot is microcontroller-based Arduino UNO board. HC-05 Bluetooth device and DC engines are connected to the microcontroller-based Arduino UNO board. The information from the smart phone is received by the HC-05 Bluetooth module and is being treated as the command to the main processing Arduino UNO board. The Arduino board after receiving the command from smart phone via Bluetooth link actuates the dc motor engine according to the command or information received.[13]

In accomplishing the undertaking, the controller is loaded with an application program written in Embedded C programing language. Related research articles in reference to realization of remote controlling of robots has been concentrated as referenced in.[1–12] Still there is need of a practicality in robotization framework, which will be anything but difficult to actualize. A case of such a financially savvy venture has been proposed here. On the off chance that you are amateur, at that point fabricating a robot (like a vehicle or an arm) is presumably one of the significant tasks to do subsequent to finding out about the nuts and bolts.

The Bluetooth module HC-05 is practically test and examined for its working and an easily available Bluetooth controller app is being provided for the mobile phones to begin transmission to the remote device.

As a continuation to that scheme, to realize the implementation of Bluetooth-controlled robot Arduino UNO is used along with some other components like wheels, motors, motor driver ICs, etc. Using Arduino UNO and some other parts, a simple mechanical vehicle is assembled which can be controlled, over Bluetooth link, through an Android app installed on user's Android phone.

Arduino is the main processing board for the commands received via Bluetooth communication link. In addition to Arduino, there are two more important modules that need to be discussed in detail for actuating the Bluetooth robot correctly. The modules are HC-05 Bluetooth and motor driver L293D. HC-05 is the Bluetooth module connected to Arduino to provide Arduino with wireless Bluetooth capability. By connecting HC-5 to Arduino, Arduino can receive and send information via Bluetooth wireless link to the connected device.

The robot implemented is provided motion using DC motors. L293D is responsible to provide the sufficient drive current to DC motors used in order to achieve the needed RPM of the motor to move the RC robot. L293D can drive two motors with speed and direction control.

All RC-controlled robots may not be the same due to differences in the circuit and structure. Discussing the circuit of the RC car discussed, let's consider the circuit or connections of HC-05. The main pins of HC-05 are Vcc, Gnd, Tx, and Rx. The Vcc, that is, +5 volt of HC-05 is connected to +5 V of power supply and Gnd to Gnd of power supply. The information flow is from the mobile phone, as a command to the robot, and no information is expected from Arduino to mobile phone. Looking into this Tx pin of the Bluetooth module, HC-05 is connected to the Rx pin of the Arduino board. The Rx pin of the Arduino depends on the argument given in the object created using the Arduino library file Software Serial. Here, pin number 2 and pin number 3 are given as the argument to the object created using the Arduino library file Software Serial. This defines the pin number 2 as Rx pin and pin number 3 as Tx. HC-05 Rx pin is left open. General input/output (GPIO) pins from 9 to 12 of Arduino are assigned to the L293D motor driver. Motor driver chip L293D has signaling pins with naming as IN1 to IN4. Motor driver chip signaling pins are connected to the Arduino GPIO pins 9–12 for signaling. As motors driver IC can drive two DC motors, it has two enable pins, which are connected to +5 V directly. Connecting enable pins of motor pins to +5 V makes both the motor driver circuit of driver chip always active and selected.

The robot so used in the present implementation used four DC motors. As one L293D chip is capable to drive only two DC motors, in order to drive all four motors two motor driver chip is used. Both the left sides of the motors are connected to the motor driver chip and the motors on the right side are connected to second motor driver chip.

For any event form mobile phone, Bluetooth module HC-05 connected to the Arduino will collect the command. For the application, simple android application Bluetooth controller is being used.

8.4.1 WORKING

The working rule is kept as basic as could be expected under the circumstances. The working rule of the circuit has been expounded with the assistance of a square graph, of the framework interconnection as appeared in Figure 8.3. As observed Figure 8.3, a DC power supply is required to run the framework. The DC power supply feeds the Microcontroller and the Bluetooth module. The Bluetooth module gets the sign sent from an android advanced mobile phone, where the application programing coded in C language is introduced. The microcontroller, along these lines, sends guidelines, which when executed, helps in the working of the engine driver (Fig. 8.4).

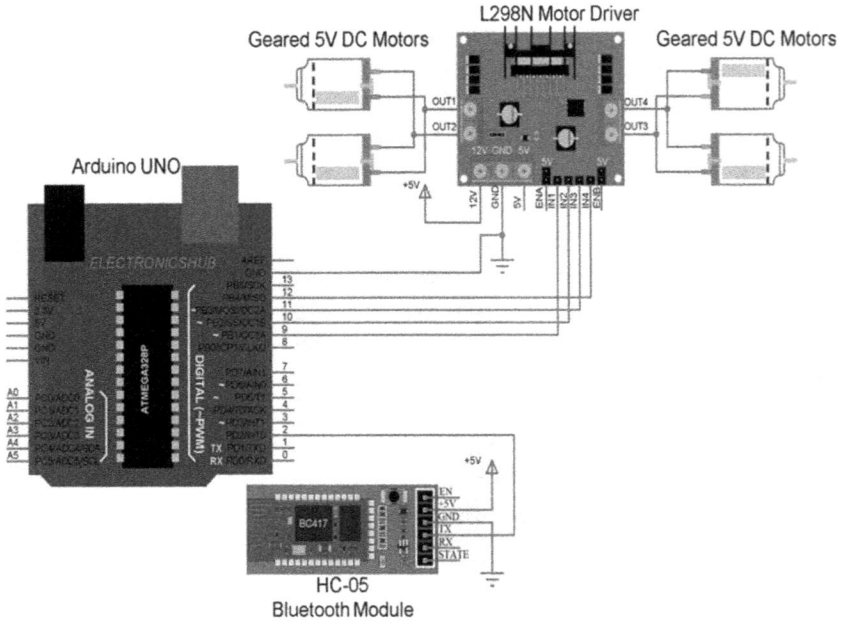

FIGURE 8.4   Circuit diagram.

The development and working of the engine can be constrained by utilizing the android based application programing. Equipment of this undertaking comprises of Arduino UNO, Bluetooth module, and an engine driver IC. The Bluetooth module is associated with the Arduino UNO board for the association with the client. Through the Bluetooth module for observing and controlling the specific engine arrives at the board and procedure as needs be and the yield of the Arduino goes to the engine driver IC and it controls the specific engine.

Right now, framework comprises of the accompanying three segments:

- Input segment
- Microcontroller segment
- Output segment

In an Android application-based Bluetooth-controlled mechanical vehicle, the client corporates the framework with an advanced mobile phone. The framework is available to be controlled in the range <15 m. In future, we would attempt to expand the range utilizing Internet of Things

(IoT).[12] At the point when client sends any information to the Arduino board then the relative pin of the Arduino goes to high state and switches ON the engine driver IC. The actuator engine moves according to the information. Here right now, client (Android application) is the information segment. This gadget is associated with the Arduino board (microcontroller segment) by the methods remotely used for examples related to Bluetooth module.

The framework would now be able to be associated with the engines (yield segment) to be controlled through remote availability. At that point through the information link, we upload the commands in the microcontroller ATMEGA 328P. These instructions help the microcontroller to interface with the Bluetooth module HC-05 and furthermore with the engine driver IC L293D. Here, the Bluetooth module goes about as a beneficiary who gets guidance from the advanced mobile phone (remote or transmitter).

At that point, the microcontroller chooses the activity for the guidance which is originating from the advanced mobile phone. The elements of the given directions are worked by the microcontroller. The guidelines are sent by the advanced cell. We can control without much of a stretch the developments of the DC engine. The Bluetooth module can work beneath the 10 m go, which we would attempt to reach out in future. Here, we are utilizing four 12 V, 200 R.P.M DC engines and a 9 V DC battery as primary force supply of this framework. Until we send any guidance to the microcontroller the engines remain in stop mode. At the point when any information is given then the engines move according to the preloaded capacities in the microcontroller (Fig. 8.5).

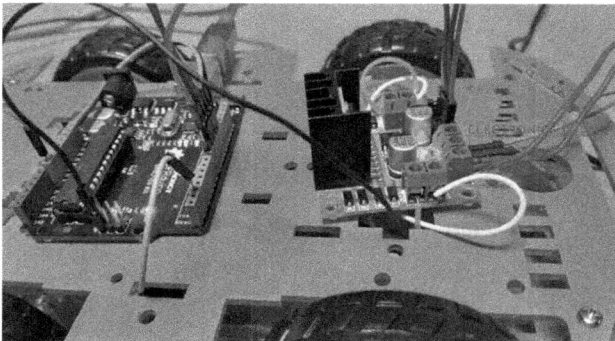

FIGURE 8.5   Representative image of hardware.

Collect the robot, make the essential associations and transfer the code to Arduino. On the off chance that you comprehended the HC-05 Bluetooth Module instructional exercise, at that point understanding the Bluetooth-Controlled Robot venture is simple (Fig. 8.6).

8.4.2   HC-05—BLUETOOTH MODULE

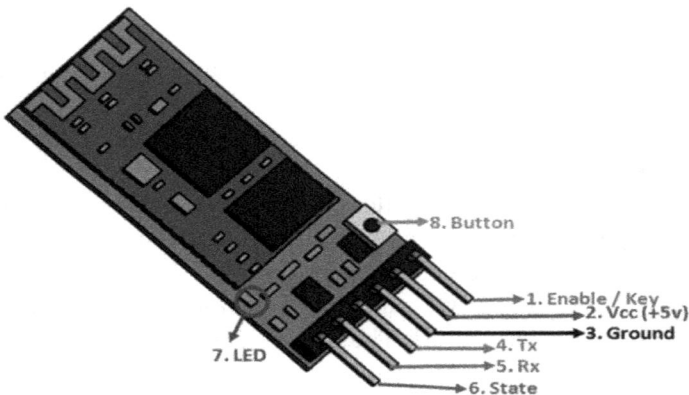

FIGURE 8.6   Hc-05 Bluetooth module.

8.4.3   PIN CONFIGURATION

Vcc: This is the power supply pin of module and is connected to +5 V.

Ground: This is also power supply pin of module and is connected to ground of power supply.

Tx—Transmitter: This is serial data transmission pin of Bluetooth module. The information received by the module via wireless Bluetooth link is being read through this pin by the Arduino.

RX—Receiver: This is serial data receiving pin of Bluetooth module. The sequence of the data to be transferred through Bluetooth is to be provided to this pin sequentially by the Arduino to HC-05 Bluetooth device.

Driven: The driven pin of Bluetooth module is connected to the on-board LED. This pin or on-board LED gives the feedback regarding the status of Bluetooth module whether Bluetooth module is in scanning mode, connected or in command mode. Flickering of on board LED once in 2 sec indicates that module has entered Command Mode, repeated Blinking of onboard LED indicated that the device is waiting for association in Data Mode while blinking fog on board LED twice in 1 sec indicate that the device has connected to another device, such as smart phone in pour case, successfully in Data Mode.

Catch: Used to control the Key/Enable pin to flip among Data and order Mode.

The Bluetooth module HC-05 comes with certain default settings which are illustrated as follows:

1. The default Bluetooth device name is "HC-05."
2. The default device password is 1234 or 0000.
3. The device HC-05 is configured as Slave by default.
4. The default operating mode for HC-05 is Data Mode.
5. The communication speed or baud rate is configured as 9600 for the HC-05 Bluetooth device.
6. The command mode baud rate for HC-05 Bluetooth module is 38400.
7. Default firmware: LINVOR.
8. HC-05 Technical Specifications.
9. Sequential Bluetooth module for Arduino and different microcontrollers.
10. Working Voltage: 4–6 V (Typically +5 V).
11. Working Current: 30 mA.
12. Range: <100 m.
13. Works good with Serial correspondence (UART) and TTL.
14. Follows IEEE 802.15.1 institutionalized convention.
15. Utilizations FHSS.
16. Can work in Master, Slave, or Master/Slave mode.
17. Can be effectively interfaced with Laptop or Mobile telephones with Bluetooth.
18. Bolstered baud rate: 9600,19200,38400,57600,115200,230400,46 0800.

As some command in the form of pressing of button from Mobile phone is generated, it is passed onto the Arduino Board wirelessly through Bluetooth Communication. The Arduino is programed accordingly to perform the operations with respective reception of the commands. For example, if "H" is passed from phone to Arduino and it is programed to switch on the LED on reception of "H," then it will.

8.5   ALGORITHM

Step 1: Assemble/connect all components.
Step 2: Install Bluetooth Android App on mobile.
Step 3: Pair the HC-05 with mobile.
Step 4: Ensure that the communication establishment between Bluetooth module and mobile.
Step 5: Provide the control signal to L293D keeping any kind of symbols for each direction (Fig. 8.7).

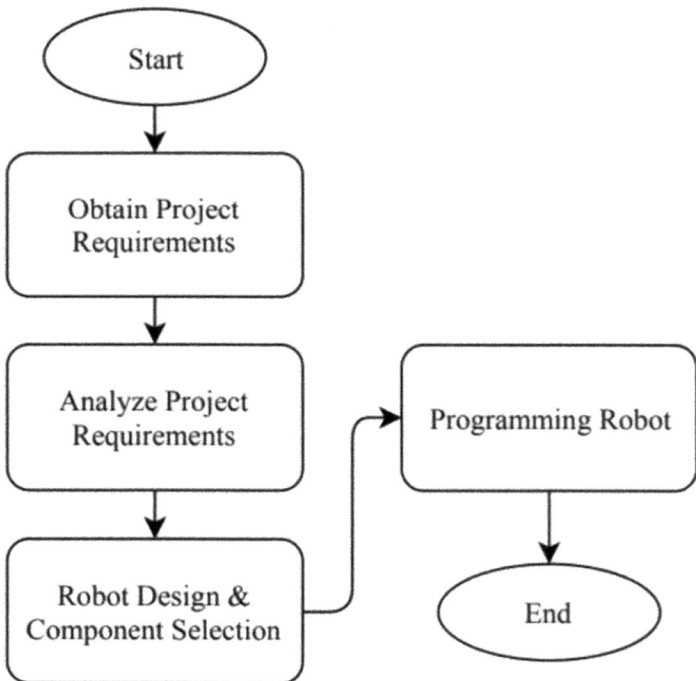

FIGURE 8.7   Flowchart.

8.6   CONCLUSION

In this chapter, we discussed about Bluetooth-controlled robot and its working, it can be implemented in various application like obstacle avoidance robot, home automation, assistive robot, military application, and healthcare application. The Bluetooth module is very helpful to provide short range communication, with help of that system can communicate with other modules. The Bluetooth-controlled robot is an alternate of human in various industries for picking and transporting of various objects where human need to involve. It can be used in military for various applications. With help of smartphones, we can operate/control and give the direction of the Bluetooth-controlled robot. There are numerous Android applications is available through which we can operate the robot.

KEYWORDS

- **Arduino**
- **Android**
- **automation**
- **bluetooth**
- **robotics**

REFERENCES

1. Akhund, T. Md. N. U. Remote Sensing IoT based Android Controlled Robot, **2019**. 10.13140/RG.2.2.31724.10882.
2. Anh, P. Q.; duc Chung, T.; Tuan, T.; Khan, M. A. Design and Development of an Obstacle Avoidance Mobile-controlled Robot. In *2019 IEEE Student Conference on Research and Development (SCOReD)* ; IEEE, Oct 2019; pp 90–94.
3. Goel, V.; Kumari, P.; Shikha, P.; Prasad, D.; Nath, V. Design of Smartphone-controlled Robot Using Bluetooth. In *Nanoelectronics, Circuits and Communication Systems* ; Springer, Singapore; pp 557–563.
4. Srikanth, P.; Ahmadhunisa, D.; Pavan, R. L.; Mounika, B.; Krishna, K. V. Hand Gesture Robot.IJRESM_V2_I3_76, 2019.
5. Taha, I. A.; Marhoon, H. M. Implementation of Controlled Robot for Fire Detection and Extinguish to Closed Areas Based on Arduino. *Telkomnika* **2018**, *16* (2), 654–664.

6. Tiwari, D. R.; Raman, D. K.; Kumar, M.; Kumar, V.; Iqbal, M. Z.; Tripathi, V. Bluetooth Contr. Floor Clean. Robot June **2018**, *1* (II). ISSN:2581–5539.
7. Lee, H. W.; Wong, C. Y. The Study of the Anti-collision System of Intelligent Wheeled Robot. In *2017 International Conference on Applied System Innovation (ICASI)*; IEEE, May 2017; pp 885–888.
8. Pinjarkar, S.; Khadpe, S.; Tavte, A.; Karpe, R. Voice Controlled Robot through Android Application. *Int. Res. J. Eng. Technol. (IRJET)* **2017**, *4* (04), 3266–3268.
9. Maity, A.; Paul, A.; Goswami, P.; Bhattacharya, A. Android Application Based Bluetooth Controlled Robotic Car. *Int. J. Intell. Inf. Syst.* **2017**, *6* (5), 62.
10. Fahmidur, R. K.; Munaim, H. M.; Tanvir, S. M.; Sayem, A. S. Internet Controlled Robot: A Simple Approach. In *2016 International Conference on Electrical, Electronics, and Optimization Techniques (ICEEOT)*; IEEE, Mar 2016; pp 1190–1194.
11. Ankit, V.; Jigar, P.; Savan, V. Obstacle Avoidance Robotic Vehicle Using Ultrasonic Sensor, Android and Bluetooth for Obstacle Detection. *Int. Res. J. Eng. Technol.* **2016**, *3*, 339–348.
12. Chanda, P.; Mukherjee, P.; SubrataModak, A. Gesture Controlled Robot Using Arduino and Android. *Int. J.* **2016**, *6* (6), 227–234.
13. Eshita, R. Z.; Barua, T.; Barua, A.; Dip, A. M. Bluetooth Based Android Controlled Robot. *Am. J. Eng. Res.* **2016**, *5*, 195–199.
14. Goon, L. H.; Isa, A. N. I. M.; Choong, C. H.; Othman, W. A. F. W. Development of Simple Automatic Floor Polisher Robot using Arduino. *Int. J. Eng. Creat. Innov.* **2019**, *1* (1), 17–23.

CHAPTER 9

Automated Aeroponics System Using IoT for Smart Agriculture

M. SURESH,* RAJESH SINGH, and ANITA GEHLOT

Lovely Professional University, Phagwara, Punjab, India

Corresponding author. E-mail: suresh.16509@lpu.co.in

ABSTRACT

The Internet of Things (IoT) has gained more coverage in the commercial as well as academic fields in recent years. Recently, IoT technologies are implemented in some other areas like intelligent cities, smart network, smart homes, autonomous cars, and some other industrial areas. Conventional agriculture is still waiting for more changes to happen particularly in IoT technology. Agricultural researchers' working along with Electronic engineers applying IoT technology in standard agriculture. The aeroponics is an effective and efficient process for farming to growing plants in a closed or open area without using any type of soil. When we design an aeroponics system along with IoT technology, many improvements as likely to happen, such as increasing plant yield, minimum usage of water, growing workforce, and minimizing growth rate. In this chapter, the new automatic aeroponics system implemented by using sensors along with IoT networking technology. This system is implemented with two components: one is a service platform and the second one is IoT devices with sensors. The service platform is providing information to mobile application from IoT devices this are using sensors in the automatic aeroponics system.

9.1   INTRODUCTION

Agriculture has an ancient past dating up to thousands of years ago. In fact, the development has been pushed by the adoption with time of the various emerging frameworks, methods, techniques, and strategies. This recruits more than a third of the world's labor force.[1] The agriculture place a major role to improve the economy for many countries and it also contributes the improvement for the underdeveloped countries to increase their economy. In fact, it also gives direction to the economic growth cycle in developing countries.

Several research articles have provided information total world's used agriculture area about 70% of the available natural resource like water each year to farming just 17% of the field. On the other hand, the overall irrigated land available is slowly decreasing due to increasingly rising the demand for food and increasing the global warming.[2,3] To put it another way, agriculture is facing new big challenges. Foote[4] told about Food and Agriculture Organization (FAO) that world's food growth will be improved by 70% to provide the fast-growing population with adequate food production and urbanization.

The rapid growth in population, in addition to decreasing the agricultural land, the global strengthening change of climate and the exaggeration of water shortages, the depletion of labor and energy crunches present immense challenges and obstacles for the agricultural area.[5,6] But the developed and developing countries will face huge water shortages and problems because of rapid industrialization and urbanization. The availability of freshwater for the agricultural field will be expected to decrease in the future.[7,8] To meet rising demands for the food, we need more input from agricultural systems. On the other side, we will suffer from food scarcity that will be the big challenge. Moreover, Qiu and coworkers[9] mentioned that Agricultural production growth is important not only for the food production to feed the population, but also for the industrial area as well.

The automatic control plant is a growing technology reviewed by Baudoin et al.,[10] like factory farms and greenhouse are the basic types of precise agriculture. In recent days, this approach is gaining popularity and grower's purpose. The system will provide a sufficient supply of food over the whole year. In this method, the plant growing throughout the year by automatically modifying and monitoring environmental factors such as temperature, CO_2 (carbon dioxide), and humidity availability of nutrients in restricted facilities.[11,12]

In addition, the method minimizes the effect on the ecosystem and maximizes the plant yield with substantial results comparing with conventional (open-area) cultivation.[13] Savvas and his team[14] intimated about the current soilless plant formation which is the most disorderly discovery at any time made in the field of automatic control plant growing method in agriculture. The method of the soilless approach referred to as plant processing methods that do not use soil by supplying water nutrient solution or artificial solid material as a cultivated medium rather than a soil. But this process of water supplies is related to aerologic and hydroponic farming. The roots of the plant are frequently rising for either approaches or intervals with or inside water nutrient mixture by creating a consistent predominance environment in an unnatural support system.[16,17]

As technology advances it can bring great benefits in all aspects of life to human beings. In particular, IoT technology growth is transforming many aspects of our existence, intelligent home, smart citified, smart grids, electric vehicles, and the industry. Intelligent farming methods are an environment that would advantage from this technology. It will offer many advantages to farmers and increase our carbon footprint by adapting the recent technological advancements for agriculture.

The Internet of Things (IoT)[18] is an innovation providing internet access and powerful data analytics tools to design a network where sensors can link physical objects to the internet. However, intelligent technology is still in its growth when it comes to use of IoT electronic equipment.

Aeroponics system cultivation is an easy and successful plant advancement in technology, not providing any kind of soil as a substance and with small amount of water. This chapter, we introduce new automatic control aeroponics system by IoT electronic equipment that will assist the farmer by improving productivity in agricultural area.

9.2 AEROPONICS

The process of growing plants either in a mist environment or in air without any use of a composite medium or soil is referred as Aeroponics. The Aeroponic culture varies from growing all traditional aquaponics, hydroponics, and in-vitro. The method is performed without a growing medium not similar to hydroponics, growing medium uses a liquid nutrient mixture and necessary minerals to support growth in plants; or aquaponics which uses fish wastage water.

FIGURE 9.1   Aeroponics.

The fundamental concept of the aeroponics method is to grow plants effect in a closed or semi-closed atmosphere nearby sprinkle the hanging roots and lower stem of the plant with an atomized or sprinkle, nutrient-rich solution of water. The crown and leaves, also referred to as the canopy, grow high above. Plant roots are isolated by the structure of the supporting plant. The closed-cell spray is often squeezed around the lower stem and placed in an aeroponics chamber opening that eliminates labor and expenditure; trellising is used to suspend vegetation and fruit weight for larger plants.

Preferably, the region remains free of disease and pests so the plants can grow better and faster than the plants which grow in the west. Because most aeroponics environments are not completely outwardly shut, pests and disease can still pose a threat. For any given plant species and cultivars, regulated environments advance plant production, growth, health, fruiting, and flowering.

Aeroponics is often paired with traditional hydroponics due to the flexibility of the root systems, and is used as an extremity "crop saver"—water sources along with nutrition backup—if the aeroponics device gets failed. The High-pressure aeroponics system is characterized by a high-pressure diaphragm pump through 20–50 μm nebula heads which supplies nutrients to the roots.

9.3 THE AEROPONIC SYSTEM

The aeroponics method is one type of the soilless farming approach, where generally the plants growing in air with addition of automatic control system, rather than substratum or soil culture. It is an air to water plant growing method where in controlled conditions below portion of plants specifically the roots are hung within growth chamber under total darkness. Outside the growth chamber, however, the top layer of the plant, they are crown portion, fruit, and leaves on expand. The artificial method (thermofoam or plastic) is typically providing support to the plant and to separate it into two portions (leaves and roots). Throughout the method, the plant roots are freely open with in the air and irrigated direct at different intervals with a small size of droplet size for the water nutrient mixture. The nutrient solution is sprayed with or without high air pressure by different atomization nozzles. This refers to the equipment and device elements that are assembled in an air culture to support plants. The aeroponics greenhouse refers to a plastic or glass structure that is managed by the atmosphere with equipment to grow plants in the mist environment or air.

The aeroponics systems feed plants with nothing but nutrient-laden water. The principle builds on that of hydroponic systems, where the roots are placed in a soilless growing medium, such as coconut coir over which nutrient-laden water is pumped periodically. Aeroponics simply dispenses with the rising medium, leaving the roots to hang in the air, where the specially built misting devices regularly puff them up.

Seeds are "planted" in aeroponics systems in bits of foam packed into tiny containers, which are exposed to light on one end and nutrient mist

on the other. The foam also maintains stem and root mass as the plants
expand.

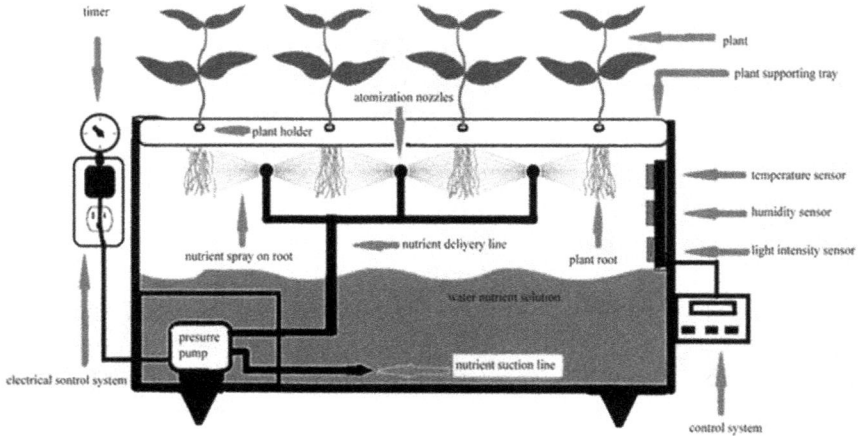

FIGURE 9.2   Aeroponic system.

9.4   IMPLEMENTATION OF AEROPONICS SYSTEM

Here, we introduce aeroponics system using two main components, one of
which is the service platform, and the second with each sensor is IoT unit.

9.5   SENSORS WITH IOT DEVICE

We executed a module to collect data from each sensor by interfacing with
Raspberry Pi3. It is one that commonly used the IoT device to connect
different sensors like (pH probe, temperature, and humidity) to analyze
data by getting from different sensors. Dependent on analyzed data, a
water pump and dosing pump controlled by Raspberry Pi3 to provide
nutrients and water to the aeroponics system.

The collected information is stored in a database for future analysis.
BCM pins in the Raspberry Pi3 connected to measure both temperature
and humidity by using DHT11 sensor, Atlas scientific pH sensor probe,
Light intensity by LDR, water quantity by water level sensor, and dosing
pumps for supply nutrients to plant.

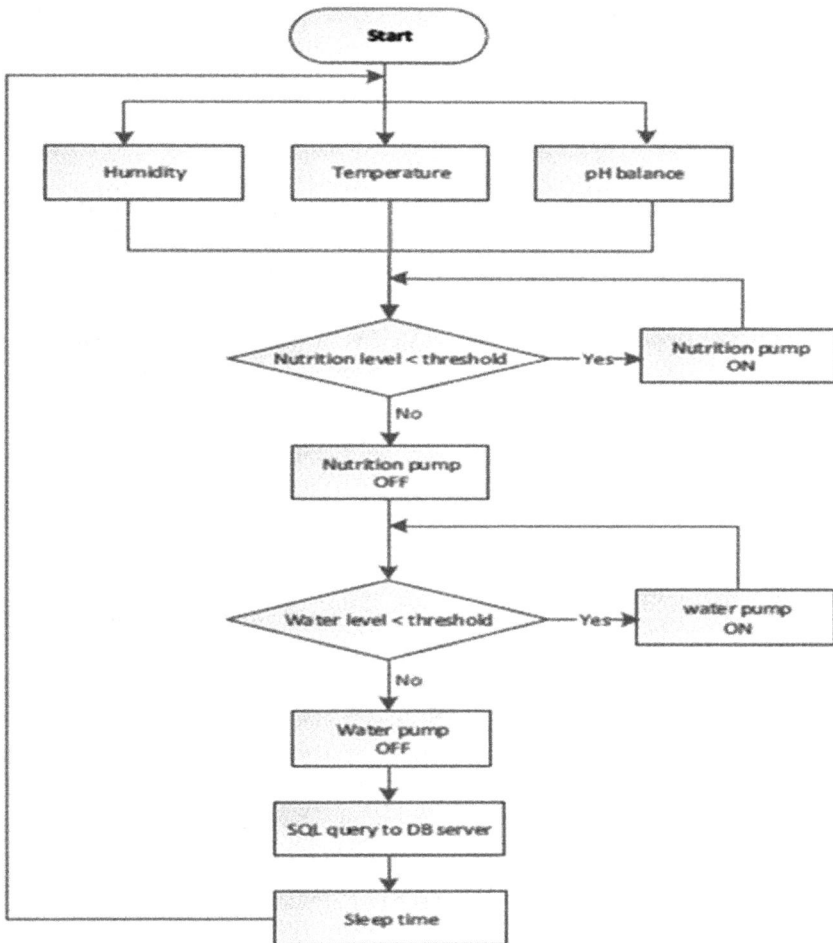

FIGURE 9.3   Flow diagram of IoT device.

The above figure provides the flow diagram of the IoT device system. Every IoT device collects data from pH probe, DHT11 Temperature, Humidity sensor at particular period of time. After analyzing sensor data, if the nutrition level of plant is less than threshold value then the controller Pi3 will active relay to turn on dosing pump to add nutrition mixture to plant in aeroponics system. If the plant nutrition mixture value reaches at threshold value it will stop the supply of nutrients in the aeroponics system.

9.6   RASPBERRY PI3

The Raspberry Pi3 is a reasonable PC based on a solitary printed circuit board. It was created in the UK by the Raspberry Pi establishment to empower the showing premise software engineering in schools and to return the enjoyable to finding out about figuring.

Specification: Raspberry Pi3 Model-B is from the third-generation which is the earliest model. As of February 2016, it replaced the Raspberry Pi2 Model B. See also the latest product from the Raspberry Pi foundation is Raspberry Pi4.

Power connector: To connect to a power socket, all Raspberry Pi3 models have a USB port (similar to found in most of the mobile phones): either USB-C for Raspberry Pi 4 or micro USB for Raspberry Pi 3, 2, and 1.

Micro SD card: Your Raspberry Pi requires an SD card to store all its Raspbian operating system and all its files. We require a Micro SD card with a capacity of **at least 8 GB**. Many sellers supply SD cards for Raspberry Pi that are already setup with Raspbian and ready to go.

A mouse and a keyboard: Using your Raspberry Pi to launch, you need a USB mouse along with a USB keyboard. Once you have set your Pi up, you can use a Bluetooth mouse and keyboard, but you will require a USB keyboard and mouse for the first setup.

A computer screen or TV: You need a screen to view the Raspbian desktop environment, and a cable to connect the screen and the Pi. The screen can be computer monitor or TV. Whether the computer has built-in microphones, the Pi will use these for sound playback.

HDMI: The Raspberry Pi3 has an HDMI output port, which is compatible with most modern TVs and computer monitors HDMI input. It is also true that many computer monitors have DVI or VGA ports.

Speakers or headphones: The ample Raspberry Pi3 models (but not Pi Zero/Zero W) had a standard type audio port like the one on your MP3 player or smartphone. If you want to, you can connect your speakers or headphones so that your Raspberry Pi3 can play sound. If your Raspberry

Pi connects to a computer that has built-in speakers, Raspberry Pi3 can play sound via these.

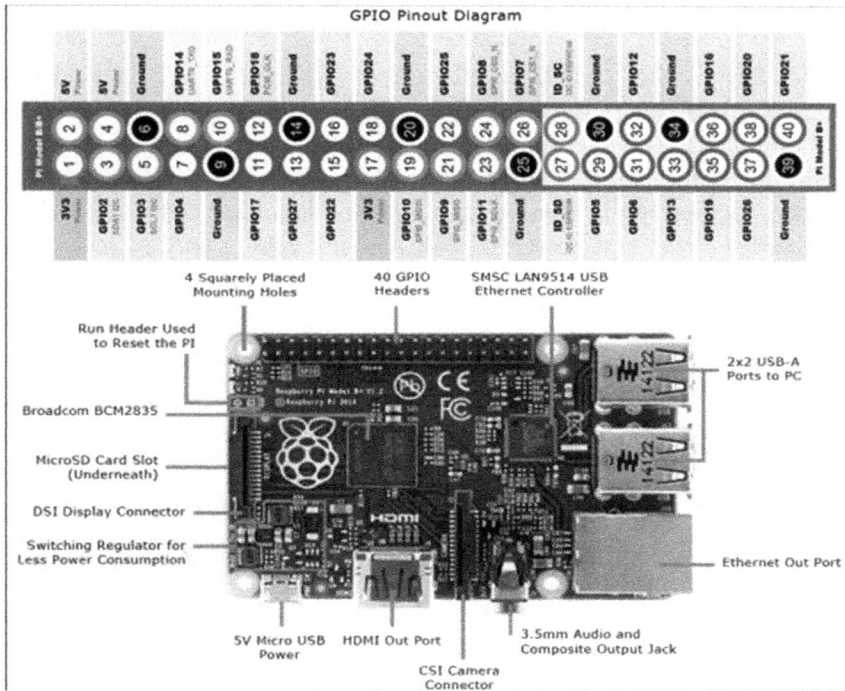

FIGURE 9.4   Raspberry Pi3.

An Ethernet cable: The large raspberry models (but not Pi Zero/ Zero W) have a standard Ethernet port to connect them to the internet; to connect Pi Zero to the internet, you need a USB-to-Ethernet adapter.

9.6.1 HOW TO INSTALL OS ON THE RASPBERRY PI3

The most commonly used operating system for Raspberry Pi is Raspbian, Debian Linux distribution spin-off that fits well on Raspberry Pi3 hardware. Raspbian is a professional along with a flexible operating system it brings all the comforts of a Desktop to your Raspberry Pi3: a command line, lots of other programs, and a browser. We can work with a Raspberry Pi3 that runs Raspbian operating system as a low-cost and powerful desktop system, or you may use it as a springboard to convert your Raspberry Pi3 into one of the countless other usable tools, Retro gaming consoles, from wireless access points. Raspbian can be installed on the Raspberry Pi3 here.

Raspbian is fairly easy to install on the Raspberry Pi3. We can load disk image Raspbian and write onto a micro SD card and then boot micro SD card onto the Pi3. We require a micro SD card for this Operating system (going for minimum of 8 GB) Desktop has a SD card slot for it, and of course a Raspberry Pi3 along with output and input devices like a screen, keyboard, mouse, and power supply. It is not only a way to install the Raspbian operating system (more on that in a moment), But learning is a useful practice since so many other Raspberry Pi operating systems can also be enabled.

9.6.2   NOOBS OS

It is another alternative method to install OS in Raspberry Pi the method mentioned here is not the only choice you have to install Raspbian OS. We may also choose to upload NOOBS, an OS installation manager that makes Raspbian, as well as several other types of operating systems, easy to install. You can purchase SD cards that come with pre-loaded NOOBS OS if you really want to make things easy.

Step 1: Download Raspbian from Pi official website

I'm going to show how to get Raspbian mounted on the Raspberry Pi3 so it's a time for us to start! First thing: jump on your machine (this is fine for both Mac and PC) and download the image from the Raspbian disk. Raspbian's latest edition is available on the Raspberry Pi Foundation website, here.

Step 2: Procedure to unzip the image file

The image file of the Raspbian is compressed, but we need to unzip the file. The file uses the format ZIP64, and you'll need to use other programs to unzip it depending on how current your built-in functionality is. Try some applications, the Raspberry Pi Foundation recommends if you have any problems:

- Linux users will use the Unzip.
- Mac users, the Unarchiver.
- Windows users want 7-Zip.

Step 3: procedure to copy disk image to micro SD card

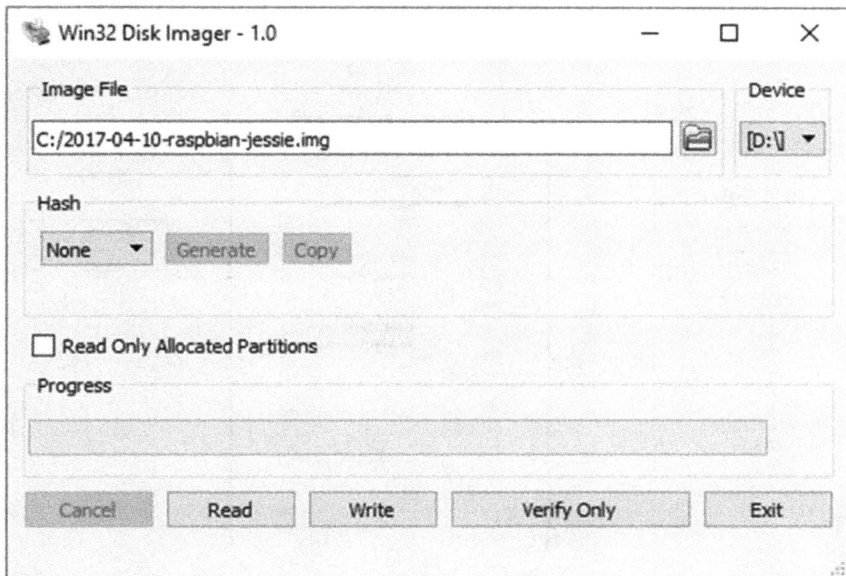

Insert micro SD card into laptop and write down the disk image to it. To do so, you'll need a fixed program:

- Linux people, Etcher—which also works on Widows and Mac—is what generally the Raspberry Pi Foundation recommends.
- Mac users can use the disk utility that's already on your computer.

- Windows users answer is Win32 Disk Imager.

For these applications, the way you write the image can be slightly different, but it's fairly self-explanatory, no matter what you're using. You will need to choose both the destination of these programs (make sure you have selected micro SD card!) and disk image (the Raspbian file that's unpacked). Select and double-check the write button then press.

Step 4: Connect micro SD card and boot into your Pi3

When disk image is transferred to the micro SD, you are good to start! Slide the sucker into your Raspberry Pi, plug in and enjoy the power source and the peripherals. The new version of Raspbian will boot right onto the desktop. Pi usernames and Raspberry passwords are your default credentials.

9.6.3   BLOCK DIAGRAM FOR PROPOSED SYSTEM

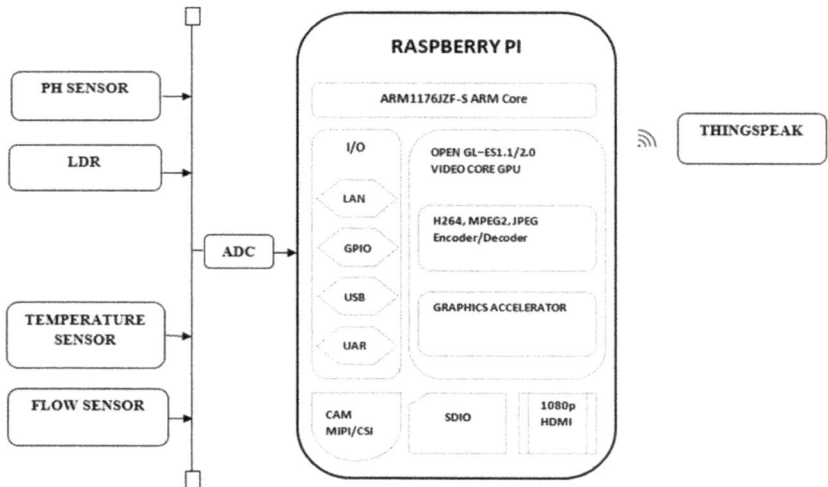

9.6.4   TEMPERATURE AND HUMIDITY (DHT11) SENSOR

The computerized temperature and moistness (DHT11) sensor are ultra-minimal effort and basic one. They utilize a capacitive dampness sensor and a thermistor to test surrounding air, and spit an advanced sign on the information pin (no requirement for simple information pins). Using it is relatively straightforward, but it desires attentive timing to capture values. The unparalleled total drawback of this sensor is that you can just get new qualities from at regular intervals, so the sensor readings can be old as long as 2 sec while utilizing our library.

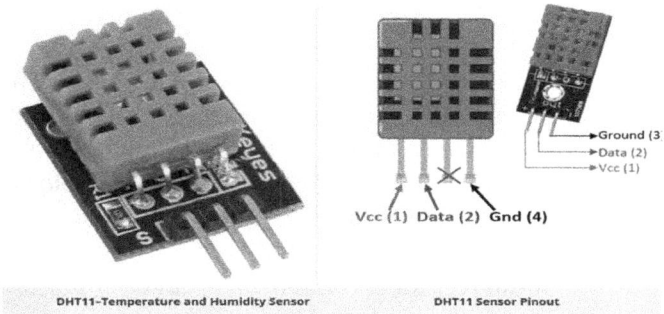

DHT11-Temperature and Humidity Sensor   DHT11 Sensor Pinout

9.6.5   PIN IDENTIFICATION AND CONFIGURATION

No:	Pin Name	Description
For DHT11 Sensor		
1	Vcc	Power supply 3.5V to 5.5V
2	Data	Outputs both Temperature and Humidity through serial Data
3	NC	No Connection and hence not used
4	Ground	Connected to the ground of the circuit
For DHT11 Sensor module		
1	Vcc	Power supply 3.5V to 5.5V
2	Data	Outputs both Temperature and Humidity through serial Data
3	Ground	Connected to the ground of the circuit

9.6.6   SPECIFICATIONS OF DHT11 SENSOR

- Temperature range is of 0–50°C

- Operating voltage is of 3.5–5 V
- Humidity range: 20–90%
- Accuracy: ±1°C & ±1%
- Operating current: 0.3 mA
- Resolution: Temperature & Humidity both are 16-bit

9.6.7   DIFFERENCE BETWEEN DHT11 MODULE AND SENSOR

The DHT11 sensor is available as a sensor module or sensor. The performances of the sensor and module are similar. There are total four pins in DHT11 sensor package among all those pins only three pins are used and one in NC (No Connection), but the sensor module available with three pin package as shown in above figure.

There is only one difference between the sensor and sensor module, which is in-built pull-up resistor and filtering capacitor in sensor module. Whenever we use sensors, we need to connect filtering capacitor and pull-up resistor for measuring temperature and humidity.

9.6.8   USAGE OF DHT11 SENSOR

DHT11 sensor is a popularly used to measure both Temperature in centigrade and humidity in percentage. The sensor built with a Negative Temperature Coefficient (NTC) for measuring temperature and an 8-bit controller for serial output data of both temperature and humidity values. The sensor is also calibrated by factory and thus easy to interface with other controllers.

The sensor measures minimum 0°C to maximum 50°C temperature and measures 20–90% humidity with an accuracy range of ±1°C and ±1%. And if we are looking to read temperature and humidity with in this range this sensor might be the right option.

9.6.9   DHT11 SENSOR WITH CONTROLLER

DHT11 Sensor is calibrated by the factory and generates data in serial and is, therefore, very simple to connect with controller. The below figure shows the connection diagram for this sensor with controller.

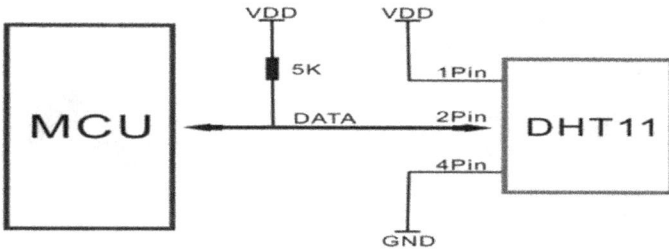

FIGURE 9.5   DHT11 sensor with controller

Applications:

- Measure humidity and temperature
- Weather station
- Automatic temperature control
- Environment monitoring

Raspberry Pi and DHT11:

The below figure shows DHT11 connected to Pi3:

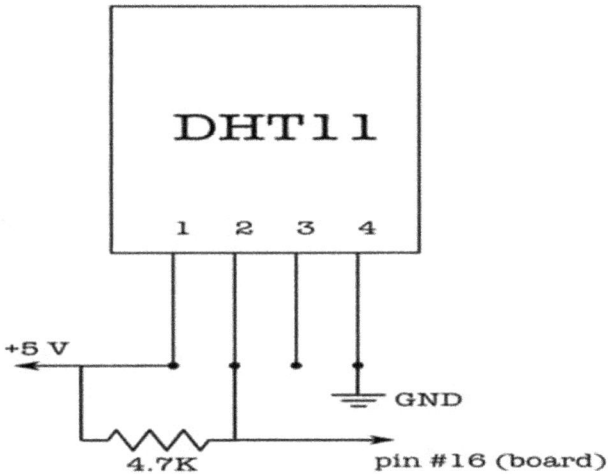

Here how it looks like on a breadboard:

FIGURE 9.6   DHT11 with Pi

We used Python language to communicate between Raspberry Pi3 with DHT11 sensor for measuring temperature and humidity, put the sensor data on ThingSpeak. Download and add Adafruit DHT Library while we are interfacing DHT11 sensor.

```
import sys
import RPi.GPIO as GPIO
import time
import Adafruit_DHT
import urllib2

def getSensorData():
RH, T = Adafruit_DHT.read_retry(Adafruit_DHT.DHT11, 23)
return (str(RH), str(T))

def main():
if len(sys.argv) < 2:
print('Usage: python tstest.py PRIVATE_KEY')
exit(0)
print ('starting...')

baseURL = 'https://api.thingspeak.com/update?api_key=%s' % sys.argv[1]

while True:
try:
RH, T = getSensorData()
```

```
f = urllib2.urlopen(baseURL + "&field1=%s&field2=%s" % (RH, T))
print f.read()
f.close()
time.sleep(15)
except:
print ('exiting.')
break
if __name__ == '__main__':
main()
```

9.6.10 THINGSPEAK

This is an "open data platform for the Internet of Things". We need to create an account after login to account create new channel with two fields Temperature and Humidity that shows what you are indicating—title, range. Then data is update in your created channel with request of an HTTP server.

9.6.11 WHAT IS PH VALUE?

The pH value is commonly used to measure which type of acids available in water like acidity and alkalinity, or to measure caustic and baseline of solution. This is usually expressed with a value between 0 and 14. The meaning pH 7 stands for neutrality. The numbers on the scale are increasing, while the numbers on the scale are decreasing with increasing acidity. In acidity or alkalinity, each unit of change reflects 10 times. The value of pH equal to the negative logarithm of the concentration of hydrogen-ion or hydrogen-ion activity as well.

The most commonly used technique of an electrochemical pH sensor is to be used to calculate the pH. The Combination pH probe sensors are a kind of electrochemical pH probe sensor which attributes both electrode and reference electrode measurement. The calculated electrode senses change the pH value, while a constant signal for differentiation is given by the reference. For viewing the millivolt signal in pH units, a high impedance system, known as a pH meter is used. Combination pH probe sensor technology can be used to construct the different types of products like pH sensors in the laboratory and in the industrial or agricultural field.

The pH sensor probes are used to measure the water alkalinity or acidity between 0 and 14 levels. As the pH-value drops below 7, the water becomes more acidic. Any number above 7 is equal to greater alkalinity. Every type of pH sensor probe works differently for measuring the water quality.

The increasing pH in the setting of an atmosphere may also be an early predictor of increased pollution. The water should be deemed high if the pH level exceeds higher than 8.5, which may cause the boilers and pipes to grow in size. As described above, you can choose from four main types of pH sensors, which contain differential sensors, combination sensors, process sensors, and laboratory sensors, each suitable for different applications.

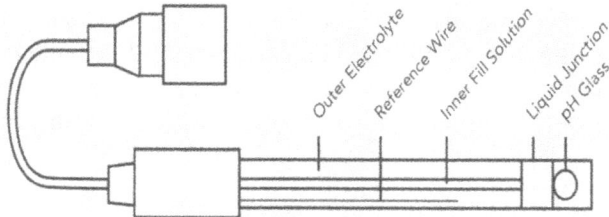

FIGURE 9.7   pH sensor probe.

9.6.12 RASPBERRY PI AND PH PROBE

Here, we showed the connected diagram

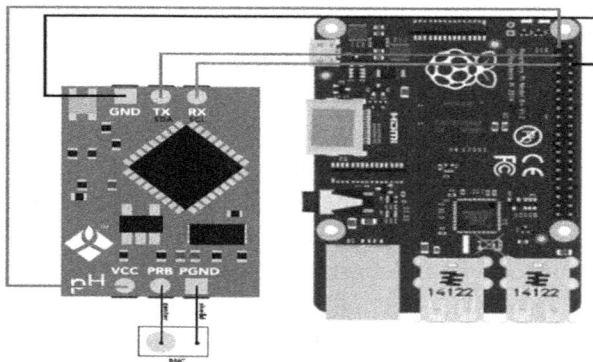

FIGURE 9.8   pH probe sensor with Pi3.

We have to make the connection of a pH probe for calculating the soil pH value. The pH probe is connected to the I2C module to Raspberry Pi. I2C module SCL and SDA are connected to the fifth and third GPIO pins of Raspberry Pi. Vcc and GND pins of I2C module connected to 5 V and GND pins of Raspberry Pi. The pH sensor function cycle just like the Flow map that shown in Figure 9.9. In this segment, it begins reading analog value when Raspberry Pi checks the link status accessible and covers the value using MCP3008. It sends a predefined message to the user end in case of link failure.

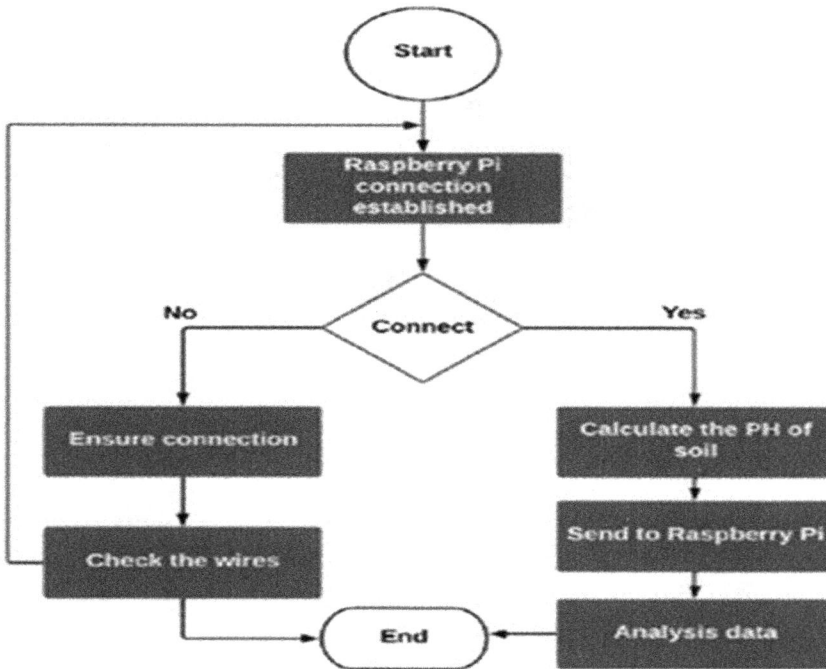

FIGURE 9.9   Flow diagram for connection pH Sensor to Pi3.

9.6.13   WATER LEVEL SENSOR

Level sensors are used to measure the amount of flowable substances. These substances include oils, slurries, granular, and powdered material. Level measurements can be done within tanks, or it can be a river or lake level. These tests may be used to assess the quantity of materials in a closed container, or the water flow in open channels.

The point level of a liquid is measured by a variety of different types of liquid level sensor. Many forms use a magnetic float, with the liquid in the container rising and dropping. When the liquid reaches certain level, and the magnet by default, a magnet switch is activated on the reed. Commonly, a transition is made to the top and bottom of the container, allowing for minimum and maximum levels of liquid to be determined. Many sensors also have a protective shield to protect the magnet from vibration or damage from direct liquid contact.

FIGURE 9.10   Water level sensor.

9.6.14   *RASPBERRY PI AND WATER LEVEL SENSOR*

The following figure demonstrates the use of a Raspberry pi3 board and a water level sensor to create a water level detector. It will detect if the container's water content exceeds the amount at issue. This device collects the relevant data via a water level sensor and sends out the data through Raspberry Pi3.

Because Raspberry Pi can only process the digital signal, to process the analog signal from the water level sensor we need to add an analog to the digital converter (ADC). It can detect smoke in the air according to the voltage value. MCP3008 is very common and highly recommended as an ADC chip.

Water level sensor is a simple and cheaper sensor that detects water quantity via an exposed line track. Quick analog signal changes from measurement of the water level to completion. The working voltage of this sensor is DC 3–5 V, and is using 3.3 V in this session. Pay more attention to the anode and else burnout your Raspberry Pi board and sensor. You can attach sensor VCC to 3.3 V, GND to 0 V.

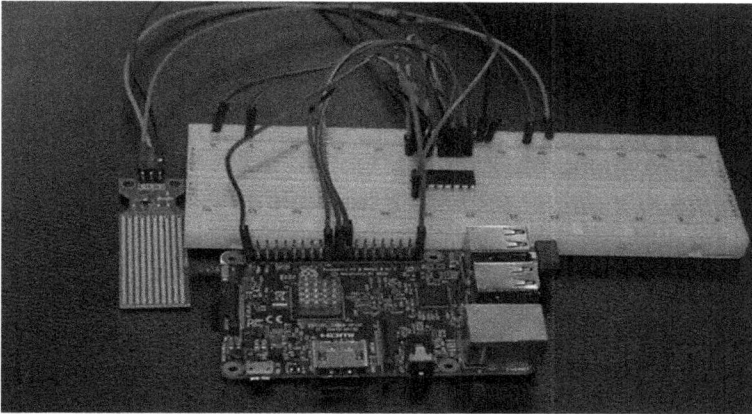

FIGURE 9.11   Water level sensor with Pi.

Basically, this is sensor use to identify the overfull of water in the field. Raspberry Pi3 sends only either "0" or "1" the binary value. We use this water flow sensor to identify the amount of water crossing the limit. The working criterion is as shown in the following Figure.

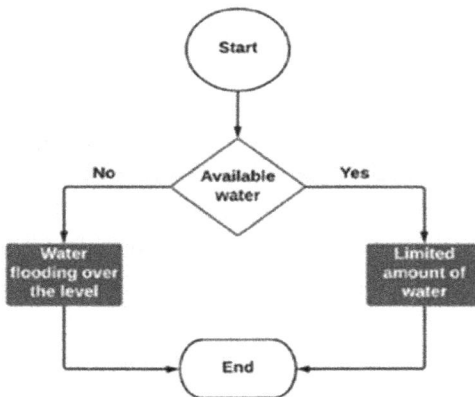

FIGURE 9.12   Flow diagram for water level sensor connecting with Pi3.

9.7 LIGHT DEPENDENT RESISTOR

The LDR (also refer as Light Dependent Resistor or photoresistor) is an apparatus who's resistivity is a part of electromagnetic radiation incident. LDRs are constructed from high-resistance, semi-conductor materials. Several distinct symbols are used to denote the LDR or a photoresistor, symbol which is commonly used is shown below figure. The symbol of arrow depicts light dropping inside.

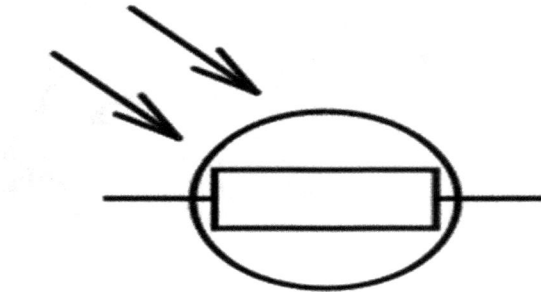

FIGURE 9.13   LDR basic symbol.

The photoresistors work on the photoconductivity principle, which is an optical effect that makes the image more conductive as the object absorbs light.

As light falls the electrons in the semiconductor material's valence band get excited to the conductive band that is when the photons land on the surface. Such photons should have more energy in the incident light than the band gap of the semiconductor material in the incident light to allow the electrons to pass from the valance band into the conductive band.

Therefore, when the light has sufficient energy reaches to the unit, the conductive band is having more and more electrons excited, resulting in a huge number of charging carriers. The consequence of this mechanism is that when the circuit is closed, more current flows to the device continuously and thus the device's resistance is said to have been lower.

Photoresistor LDRs are light-reliant gadgets, when light drops on them, decline obstruction and they increment in obscurity. It has high opposition if a light-needy resistor is kept in obscurity. It has high opposition if a light-reliant resistor is kept in obscurity. This obstruction is called

Darkness' Resistance. For a given LDR, the following figure indicates the resistance vs. illumination curve.

Photocells are nonlinear devices, or LDR's. Their sensitivity varies with the light incident wavelength upon them. Some photocells may not be able to respond to any given range of wavelengths. Different cells, depending up on the type of material used, have different spectral response curves.

Upon the incident of light on a photocell, the increase in resistance typically takes around 8–12 ms, whereas after the light is withdrawn, rising back to its initial value takes 1 or 2 sec for the resistance. This phenomenon is called the rate of recovery of resistance. Use these features in audio compressors.

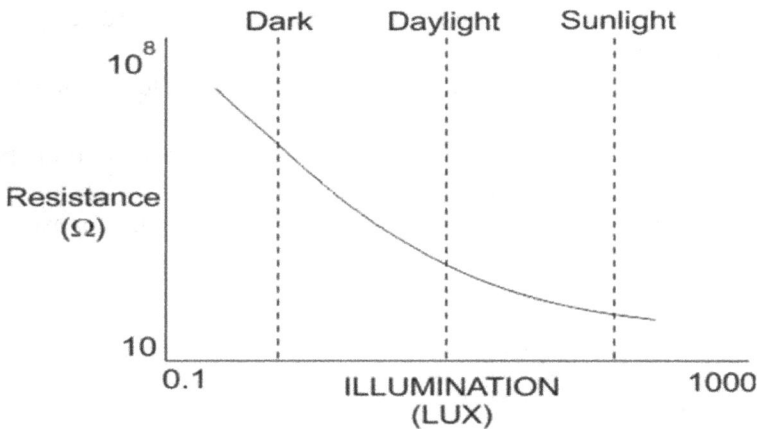

FIGURE 9.14   Resistance vs. illumination.

Photoresistors (LDRs) are divided into two groups based on the type of materials used to create them. The two types of photoresistors comprise:

1. **Intrinsic photoresistors** (undoped semiconductor): These are made of pure materials like silicon and germanium. As photons with sufficient energy fall on them, increasing the number of charging carriers, electrons get excited from valance band to conductive band.

2. **Extrinsic photoresistors:** They are doped products of semiconductors with impurities called dopants. Such dopants produce new

energy bands which are packed with electrons above the valence band. This reduces the band gap and takes less energy to arouse them. Long wavelengths are commonly used for extrinsic photographic resistors.

The LDR structure is a light-touchy material that is stored on a protecting substrate, for example, clay. The material is saved in a crisscross example to get the ideal opposition and force level. This crisscross zone isolates the metal-installed zones into two unmistakable areas.

9.7.1 RASPBERRY PI3 WITH LDR

The following steps need to follow while interfacing LDR with pi3.

1. Initially, we wire pin no. 1 (3.3v) on the breadboard refer to the positive rail.
2. We wire pin no. 6 (GND) on the breadboard refer to the ground rail.
3. We put the LDR onto the breadboard and take a wire to the positive rail from one end.
4. Spot a wire driving back to the Raspberry Pi, on the opposite side of the LDR sensor, place on pin no. 7.

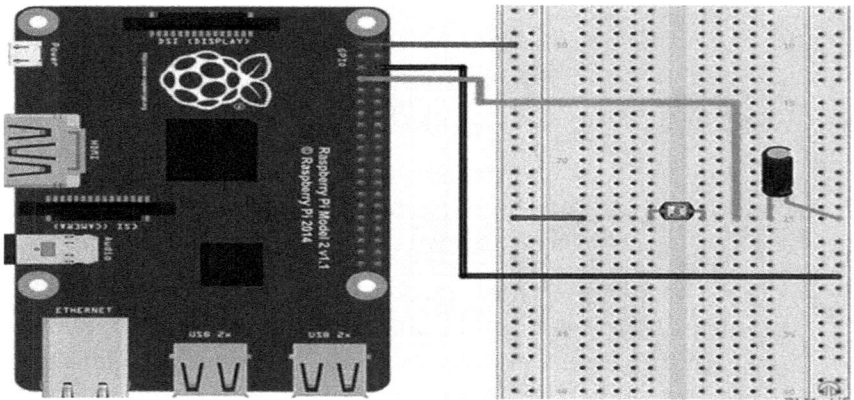

FIGURE 9.15 LDR with Raspberry Pi3.

9.7.2   A 12V PERISTATIC PUMP

With this very cool little pump you transfer the fluid safely from here to there. Unlike other types of liquid pumps, this is a peristaltic form—the pump that squishes the liquid containing silicone tubing instead of directly impelling it. The Revolt? The pump never touches the fluid making it a perfect option for any food/drink/sterile-dependent pumping like making drink-bots.

The pump is basically a geared DC motor, and there is a lot of torque to it. Within the pump is a "clover" roller pattern. The clover pushes the tube to pass the fuel, while the motor spins. The pump does not need to be ground and can be conveniently self-printed with half a meter of water by PWM the engine you can accelerate or slow down the flow rate and if you mount the engine then it will push the fluid in the other direction. Works perfectly with either the power transistor (basic on/off) or L293D motor driver chip.

The pump comes with a bit of silicone tubing, and two 1/2 meter parts with barbed connectors are included. Nevertheless, the silicone tubing is not clean and maybe sticky on the way to you. Upon using the tubing needs to be sterilized! In addition, the tubing offered is not compatible with FDA or USDA and is intended solely for basic pump testing. Check McMaster-Carr if you need to buy FDA/USDA-complaint tubing for use in your food project.

9.7.3   TECHNICAL DETAILS

- Working Temperature: 0–40°C
- Uses about. 4 mm outer diameter, 2 mm internal silicone tube, the size of the pump tube has changed on us, so please weigh the tube that comes to check with your pump.
- Motor current: 200–300mA
- Motor voltage: 12V DC
- Weight: 200 grams
- Flow rate: up to 100 mL/min
- Mounting holes: 50 mm center-to-center distance, 2.7 mm diameter,
- Dimensions: 27 mm diameter motor, 72 mm total length

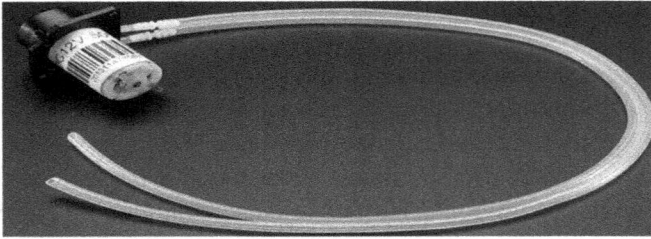

FIGURE 9.16   Peristaltic liquid pump.

9.7.4   RELAY

A relay is an electromagnetic switch that is used with a low power signal to turn on and off a circuit, or when several circuits have to be operated by one signal. We know most high-end industrial application devices are fitted with relays for efficient operation. Relays are simple switches that are operated both electrically and mechanically.

Relays consist of an electromagnet and contacts as well. The method of switching is applied with the aid of the electromagnet. Many operating standards are also in place for its function. But they vary in the way they apply. Most of the tools have relay functionality. It is an electromagnetic relay rounded by an iron core with a wire coil.

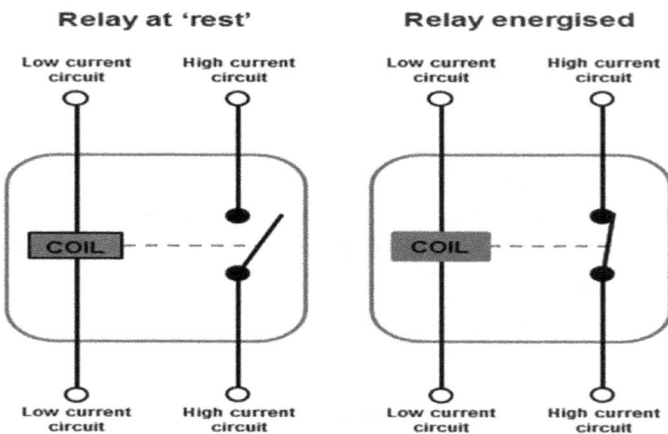

FIGURE 9.17   Relay.

9.7.5   RASPBERRY PI AND PUMP

In this Raspberry Pi and pump interfacing, we're going to automate the whole nutrition of plants with a Raspberry Pi linked to a Temperature, Humidity, pH value, and an add-on board that is connected to a pump.

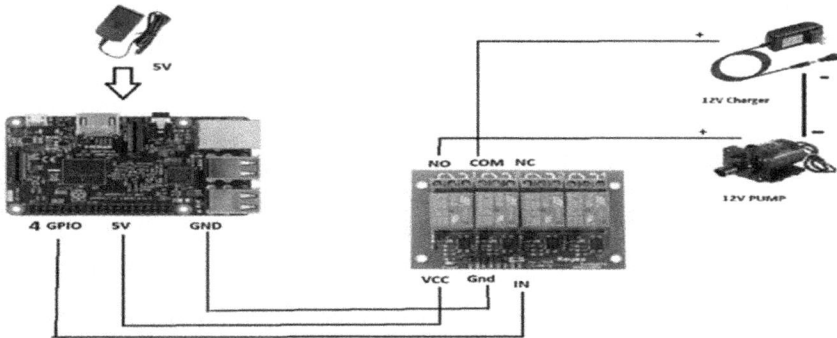

FIGURE 9.18   Raspberry with pump.

The following steps we need to follow while we are interfacing pump with Pi3

1. Wire relay Vcc to voltage 3.3 V from Raspberry Pi3.
2. Wire relay GND to Ground from Raspberry Pi3.
3. Connect relay module to any GPIO pin (in this module I used pin 7[GPIO4]).
4. Execute the below program provided below.

```
import RPi.GPIO as X
import time
GPIO.setwarnings(False)
GPIO.setmode(X.BOARD)
GPIO.setup(7,X.OUT)
from Tkinter import*
myGui=Tk()
while True:
def lampon():
GPIO.output(7,True)
def lampoff():
```

```
GPIO.output(7,False)
myGui.title("Hello")
myGui.geometry("200x350+200+200")
while True:
myButton1=Button(text='on',fg='black',bg='green',command=lampon).
pack()
myButton2=Button(text='off',fg='black',bg='green',command=lam
poff).pack()
myGui.mainloop()
```

9.7.6   MCP3008

The MCP3008 is a 10-bit ADC (analog to digital converter), having eight channels analog inputs to read from Raspberry Pi3, as there is no inbuilt ADC in Pi. This IC is a significant choice if we want to read any type of analog signals like a Temperature and Humidity sensor (DHT11). If there is a need of more precision or features, we could use advanced series of ADC like ADS1x115.

Raspberry Pi's SPI serial communication is used to connect with MCP3008. We have the option to use either any four general purpose input/output pins along with the software SPI or hardware SPI bus to communicate with MCP3008. This regard software SPI is more versatile as the programmer has the option to work with the Pi3 of BCM or GPIO pins, whereas hardware Serial Peripheral Interface (SPI) is a little faster but less flexible as it can be used to work with specific pins of Pi. Another reason to proceed with software SPI is it's easiness to setup with Pi3. You need to place the MCP3008 IC into a breadboard before you can attach the Pi3 to the chip first.

9.7.7   MCP3008 FEATURES

- 8-channel ADC IC with 10-bit resolution and serial SPI interface communication protocol.
- It has programmable analog inputs which can be configured in either single-ended or pseudo-differential modes.
- Operate over a range of 2.7–5V.

- The chip employs Successive Approximation (SAR) architecture for ADC conversion.
- The sampling rate is 200 ksps for 5 V and 75 ksps for 2.7 V, respectively.
- It is based on low power CMOS technology.
- The industrial temperature range for this chip is −40°C to +85°C.
- It has a standby current of 5 nA and typical active currents of 320 μA.

9.7.8   WHERE TO USE IT?

- There are some devices like Raspberry Pi which do not have hardware for analog to digital converter and, therefore, they cannot read analog inputs. So, you need a circuit for this conversion. For such devices, you can use the MCP3008 chip. This chip uses an SPI interface for communication. In Raspberry Pi, only four GPIO pins are required. So, you can get eight additional analog inputs by using this chip.
- Sensors use analog outputs. Therefore, many devices need an ADC converter to read these outputs. The MCP3008 can be used for converting these analog signals into digital signals.

9.7.9   HOW TO USE MCP3008?

- It consists of a famous SAR ADC architecture technology that contains a built-in sample and holds a capacitor. This design performs inspecting with an example/hold capacitor for 1.5 clock cycles on the principal rising edge of the clock cycle. After that ADC produces 10-bit digital output depending on the charge value on S/H capacitor.
- The communication is initiated with the MCP3008 device is accomplished by bringing the CS line low. On the first clock signal (when CS is low and DIN is high), the first bit received will constitute a start bit. This start bit is followed by the SGL/DIFF bit which determines the mode of conversion either single-ended or differential. After that, the following three bits which are D0, D1, and D2 are utilized to choose the channel. On the fourth rising edge of the

clock, after the beginning piece has been gotten, the examining of simple data sources will be begun.

9.7.10   PIN DIAGRAM

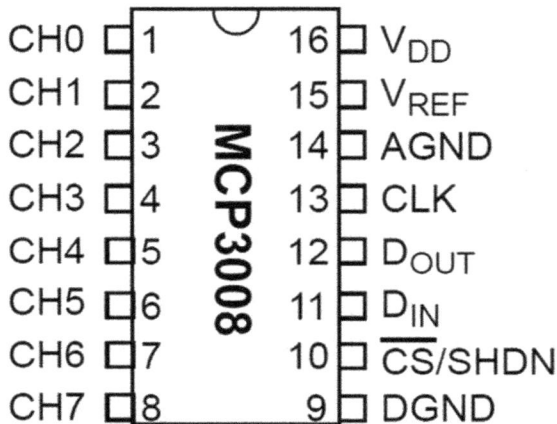

FIGURE 9.19   MCP3008 pin diagram.

9.7.11   PIN DESCRIPTION

The pins description of MCP3008 chip is given below:

Pin#01 to 08: CH0 to CH7: These are the analog inputs for channel 0–7. These channels can be configured as four single-ended inputs or two pseudo-differential pairs. In pseudo-differential mode, each channel pair is programed as the IN+ and IN− inputs by sending a serial command string.

Pin#09: DGND: This is the digital ground pin which is linked internally to the digital circuitry of chip.

Pin#10: /SHDN: It is a Chip Select pin. This pin is utilized to start correspondence with the gadget by interfacing it to a low rationale level. If it is already low, then it should be pulled to high and then low for initiating communication. When it is pulled to high logic, it will end a conversion.

Pin#11: Din: This is the input pin for serial data.

Pin#12: Dout: It is the serial data output used for SPI communication. On every falling edge of clock signal, data will change and this converted data are shifted out on this pin.

Pin#13: CLK: It is a serial clock signal used to initiate a conversion and sends each bit out as conversion takes place.

Pin#14: AGND: It is the analog ground pin which is connected internally with analog circuitry. It is connected to the reference voltage.

Pin#15: Vref: It is connected to the reference voltage and is used to determine the range of analog voltage.

Pin#16: VDD: It is the connection for applying a positive voltage to the circuit.

9.7.12   SOFTWARE SPI

To connect Raspberry Pi to MCP3008 Analog—Digital convertor we make the connections shown below for software SPI.

- Raspberry Pi3 3.3V pin to MCP3008 pin VREF
- Raspberry Pi3 3.3V pin to MCP3008 pin VDD
- Raspberry Pi3 GND pin to MCP3008 pin DGND
- Raspberry Pi3 GND pin to MCP3008 pin AGND
- Raspberry Pi3 pin 23 to MCP3008 pin DOUT
- Raspberry Pi3 pin 18 to MCP3008 pin CLK
- Raspberry Pi3 pin 25 to MCP3008 pin CS/SHDN
- Raspberry Pi3 pin 24 to MCP3008 pin DIN

You need to modify pins according to your program.

9.7.13   HARDWARE SPI

To use SPI hardware first makes sure to use the raspi-config tool to allow SPI.

The following connections show how Raspberry Pi connects to the MCP3008:

- Raspberry Pi3 3.3V pin to MCP3008 pin VREF
- Raspberry Pi3 3.3V pin to MCP3008 pin VDD

- Raspberry Pi3 GND pin to MCP3008 pin DGND
- Raspberry Pi3 GND pin to MCP3008 pin AGND
- Raspberry Pi3 MISO pin to MCP3008 pin DOUT
- Raspberry Pi3 SCLK pin to MCP3008 pin CLK
- Raspberry Pi3 CE0 pin to MCP3008 pin CS/SHDN
- Raspberry Pi3 MOSI pin to MCP3008 pin DIN

KEYWORDS

- **smart farming**
- **sensors**
- **automated aeroponic system**
- **IoT**

REFERENCES

1. James, J.; Maheshwar, M. P. Plant Growth Monitoring System, with Dynamic User-interface. In *2016 IEEE Region 10 Humanitarian Technology Conference (R10-HTC)*; Agra, India, December 2016; pp 1–5.
2. Pimentel, D.; Berger, B.; Filiberto, D. et al., Water Resources: Agricultural and Environmental Issues, *Bioscience* **2004,** *54* (10), 909–918.
3. Taher Kahil, M.; Albiac, J.; Dinar, A. et al., Improving the Performance of Water Policies: Evidence from Drought in Spain. *Water* **2016,** *8* (2), 34.
4. Foote, W. *To Feed the World in 2050, We Need to View Small-Scale Farming as a Business*; Skoll World Forum: Oxford, 2015.
5. Doknić, V. *Internet of Things Greenhouse Monitoring and Automation System. Internet of Things: Smart Devices, Processes, Services*; 2014. http://193.40.244.77/idu0330/wpcon ent/uploads/2015/09/140605_Internet-of-Things_Vesna- Doknic.pdf.
6. Großkinsky, D. K.; Svensgaard, J.; Christensen, S.; Roitsch, T. Plant Phenomics and the Need for Physiological Phenotyping across Scales to Narrow the Genotype-tophenotype Knowledge Gap. *J. Exp. Bot.* **2015,** *66* (18), 5429–5440.
7. Playán, E.; Mateos, L. Modernization and Optimization of Irrigation Systems to Increase Water Productivity. *Agric. Water Manage.* **2006,** *80* (1–3), 100–116.
8. Levidow, L.; Zaccaria, D.; Maia, R.; Vivas, E.; Todorovic, M.; Scardigno, A. Improving Water-efficient Irrigation: Prospects and Difficulties of Innovative Practices. *Agric. Water Manage.* **2014,** *146*, 84–94.
9. Qiu, R.; Wei, S.; Zhang, M. et al. Sensors for Measuring Plant Phenotyping: A Review. *Int. J. Agric. Biol. Eng.* **2018,** *11* (2), 1–17.

10. Baudoin, W.; Nono-Womdim, R.; Lutaladio, N. et al. *Good Agricultural Practices for Greenhouse Vegetable Crops: Principles for Mediterranean Climate Areas (No. 217)*; Food and Agriculture Organization of the United Nations: Rome, 2013.

11. Lee, M.; Yoe, H. Analysis of Environmental Stress Factors Using an Artificial Growth System and Plant Fitness Optimization. *BioMed Res. Int.* **2015,** *2015*, Article ID 292543, 6 pages.

12. Moon, S. M.; Kwon, S. Y.; Lim, J. H. Minimization of Temperature Ranges between the Top and Bottom of an Air Flow Controlling Device through Hybrid Control in a Plant Factory, *Sci. World J.* **2014,** *2014*, Article ID 801590, 7 pages.

13. Stanghellini, C. Horticultural Production in Greenhouses: Efficient Use of Water. *Acta Horticulturae* **2014,** *1034*, 25–32.

14. Savvas, D.; Gianquinto, G.; Tuzel, Y.; Gruda, N. Soilless Culture. In *Good Agricultural Practices for Greenhouse Vegetable Crops: Principles for Mediterranean Climate Areas (No. 217)*; Baudoin, W., Nono-Womdim, R., Lutaladio, N., Hodder, A., Castilla, N., Leonardi, C., Pascale, S., Qaryouti, M., Journal of Sensors 13, Eds.; Food and Agriculture Organization of The United Nations, Rome, 2013; pp 303–354.

15. Lakhiar, I. A.; Liu, X.; Wang, G.; Gao, J. Experimental Study of Ultrasonic Atomizer Effects on Values of EC and pH of Nutrient Solution. *Int. J. Agric. Biol. Eng.* **2018,** *11* (5), 59–64.

16. Beibel, J. P. *Hydroponics -The Science of Growing Crops without Soil*; Department of Agriculture: Tallahassee. Bulletin, 1960.

17. Reyes, J. L.; Montoya, R.; Ledesma, C.; Ramírez, R. Development of an Aeroponic System for Vegetable Production. *Acta Horticulturae* **2012,** *947*, 153–156.

18. Ashton, K. That "Internet of Things" Thing. *RFiD J.* July **2009,** *22* (7), 97–114.

CHAPTER 10

Requirements of Applications of Wireless Sensor Networks for the Internet of Things

RAKESH KUMAR SAINI[1*], MOHIT KUMAR SAINI[2], and
RAVINDRA SHARMA[3]

[1]School of Computing, DIT University, Dehradun, Uttarakhand, India

[2]Department of Computer Science, Doon Business School, Dehradun, Uttarakhand, India

[3]Swami Rama Himalayan University, Dehradun, Uttarakhand, India

*Corresponding author. E-mail: rakeshcool2008@gmail.com

ABSTRACT

Wireless sensor networks (WSNs) are used inside the Internet of Things (IoT) so as to enhance the productiveness and performance of present and prospective manufacturing industries. The IoT covers a large range of industries and uses times that scale from an unmarried restrained tool as tons as massive go-platform deployments of embedded generation and cloud structures connecting in real-time. WSNs are one of the foremost pillars for a lot IoT packages. WSNs and IoT devices are proliferated in lots of domains such as crucial infrastructures including electricity, transportation, and manufacturing. Consequently, the maximum of the everyday Operations now rely upon the information coming from Wi-Fi sensors or IoT devices and their actions. The knowledge of IoT is particularly appreciated or people with incapacities, as IoT machineries container provision social accomplishments at greater measure comparable construction or civilization. We present a survey wherein wireless sensor packages are used to manipulate IoT home equipment in a clever building. Encountered issues are highlighted and suitable answers are mentioned. WSNs are characterized by excessive heterogeneity due to the fact there are numerous distinctive proprietary and non-proprietary solutions. This huge variety of era has no longer on time new deployments and integration with gift sensor networks. The contemporary fashion, however, is to transport some distance from copyrighted and locked necessities, to consist of IP-based totally sensor networks the usage of the evolving desired 6LoWPAN/IPv6. This permits indigenous connectivity among WSN and Internet, allowing smart gadgets to take part to the IoT. The Internet is efficiently transferring after an Internet of persons concerning an IoT. By 2021, it is predictable to have 70 billion effects associated with the Internet. Though, such a movement encourages a robust equal of complication when management interoperability between the dissimilar Internet belongings, for example, RFIDs (Radio Regularity Documentation), moveable handheld strategies, and wireless instruments. This chapter describes applications requirements of WSNs applications for IoT. In this chapter, we evaluate different processes to integrate wireless sensor networks interested in the Internet.

10.1 INTRODUCTION

The IoT is developing rapidly and in destiny human beings each day needs going to rely upon the net. It is no longer simply connecting computers and smartphones anymore. Multiple devices that we use in daily life want the net to serve humans. The primary assistances of this broadsheet may be concise by way of surveys: We observe Wireless Sensor Networks (WSNs) and the Internet holistically, consistent with the imagination and perception in which WSNs may be a portion of an Internet of Things (IoT). A wearable scheme like no one of a kind, the association supervisor armband intelligence vitality resounding so that you can accomplish any device connected to the IoT business enterprise objective sideways along with your venture or moves. Cloud-based human capital management solutions, insight-based analytics, and democratized dashboards will enable the professionals to create learning, collaborative, and interactive talent scape for an agile organization.[26] The armband is prepared with conductors to differentiate energy diversion and perceive discount and relaxation of them while the hand is in motion. These moves are then redirected to software on the backend that decodes and translates them into instructions and executes the movement. Imagine being at the shoes of Tony Stark and unconditionally indicating at your tablet's demonstration to open and near apps from nearby. The result is that no single structure will suit most of these regions and the requirements each region brings. However, a modular scalable structure that helps adding or subtracting abilities, as well as supporting many requirements across a huge form of these use instances is inherently useful and treasured.[3,4] It affords a place to begin for architects trying to create IoT solutions in addition to a robust foundation for additional improvement. This includes many exceptional structures, together with Internet related cars Wearable devices which include fitness and health tracking devices, watches, and even human implanted devices; smart meters and clever gadgets; home automation systems and lights controls; smartphones which can be more and more getting used to measure the arena around them; and WSNs that degree weather, flood defenses, tides, and more. The growth of the wide variety and style of devices which might be accumulating facts is rather speedy. A look at by using Cisco estimates that the range of Internet-connected gadgets overtook the human populace in 2010, and that there might be 50 billion Internet-related gadgets by 2020.The emergent knowledge of IoT is to permit in dependent conversation of convenient statistics among imperceptibly surrounded dissimilar absolutely recognizable actual creation strategies everywhere us, powered by the foremost know-hows similar Wireless-Occurrence Vulnerable of identity and WSNs which are detected by the device approaches and additional preserved for conclusion creation, on the foundation of which an automatic act is achieved. IoT is a situation of linked bodily items which are accessible thru the internet. It is the network of bodily gadgets which could talk, sense, or engage with their internal states or the outside environments. This paper survey the makes use of Wi-Fi sensor networks programs for IoT.[5,6] The effects to be associated with the Internet mostly differ in expressions of appearances. These varieties after actual minor and standing strategies (e.g., RFIDs) to big and moveable strategies (e.g., automobiles). Such heterogeneity encourages complication and specifies the occurrence of a progressive middleware that canister cover this heterogeneity and encourage limpidity. IoT will participate in irritating set of solicitations into the Internet, for example, computerization, climate detecting, and Clever Networks. The concluding is one of the further most encouraging IoT solicitations. IoT is particularly appreciated or societies with incapacities, as IoT machineries container provision social accomplishments at superior measure like structure or civilization, as the strategies can commonly collaborate to entertainment as an entire scheme. WSNs are combined into the IoT where sensor nodes intersection the Internet energetically, and use it to cooperate and achieve their responsibilities. Sensor Networks are approximately useful in the IoT in instruction to enhance the efficiency and competence of current and potential industrial. In specific, a part of attention that disquiets the use of Wireless Sensor Networks in IoT is the thought of sensor network virtualization and overlap systems. Together system virtualization and overlap systems are measured modern-day since they afford the capability to produce facilities and requests at the advantage of current computer-generated nets lacking altering the fundamental substructure. The IoT is a new and rising investigation part regarding the request of the IoT and connected skills such as WSNs in directive to expand the efficiency of numerous manufacturing procedures and schemes. With the occurrence of the IoT and Replicated Somatic System ideas, WSNs have instigated to develop additional cooperating and shareable schemes. Source distribution among diverse WSNs delivers important benefits in positions of setting up price and period, particularly for large IoT schemes such as smart city and smart energy. The supply distribution in WSNs is typically approved out in two dissimilar conducts. Those are WSNs virtualization and middleware-based server organizations training.

10.2   WIRELESS SENSOR NETWORKS

In outdated uttered communique systems, the Open Systems Interconnection (OSI) layered building has been kind of surveyed and has attended many public services constructions nicely in the past; but, developing wireless networks of these days are meaningfully tough this association attitude. The blanketed production describes a mountain of procedure layers wherein every layer causes inner its attractively distinct article and margin, and hereafter permitting rules to the essential expertise at each layer misplaced applying the need to occupation the general scheme building.

This technique has been an achievement in its potential to offer modularity, clearness, and regulation personal the chain line structures but is in all likelihood deceptive within the Wi-Fi structures location.[7,8] Though

WSNs, quit to end with cellular networks, Wi-Fi network function networks (WLANs), cellular statement-hoc networks (MANETs), and WSNs are extensively distinct in positions of their plans and structure, a common topic in these varieties of networks is the use of the wireless channel for conversation. Different the twine line networks, the Wi-Fi channel has various unmistakable originalities that want to be taken decrease than mirrored image at the same time as planning wireless networks.[9] WSNs may have a position to play for some of the navy commitments besides with enemy motion detection and stress tracking. WSNs may be implemented with the aid of the merchant marine for a number of functions composed with staring at modern awareness in distant off districts and electricity safekeeping. Being geared up with suitable sensors, these networks can stand discovery of player indication, evidence of the identity of buddy weight, and investigation of their indication and improvement. The attentiveness of this text is at the competitive substances for plastic Wi-Fi sensor networks.[1,3] WSNs includes spatially allotted independent sensors to show bodily or environmental situations, collectively with temperature, sound, vibration, stress, movement or pollution and to cooperatively skip their numbers via the network to a top region.[10,11]

These are comparable to wireless ad hoc networks in the intelligence that they trust on wireless connectivity and unprompted development of systems so that instrument data can be conveyed wirelessly. WSNs are spatially scattered independent devices to observe somatic or conservational circumstances, such as hotness, complete, heaviness, etc., and to accommodatingly permit their data through the network to a central location. The more modern networks are bi-directional, also enabling control of sensor activity. The development of WSNs was motivated by military applications such as battlefield surveillance; today such networks are used in many industrial and consumer applications, such as industrial process monitoring and control, machine health monitoring, and so on.

10.3   APPLICATIONS OF WSN

There are approximately solicitations of WSNs that are used for IoT. WSNs have expanded significant attractiveness outstanding to their tractability in resolving difficulties in dissimilar solicitation fields and have the possibility to modify our exist in many dissimilar conducts. Essentially, these solicitations are used for the IoT:

10.3.1   MILITARY

WSNs will consume a character to production for a number of military resolutions such as enemy movement discovery and strength pursuing. WSNs can be used by the armed for a number of resolutions such as viewing innovative movement in isolated areas and strength safety. Being equipped with appropriate sensors, these networks scan support detection of opponent undertaking, certification of adversary strength and examination of their responsibility and development. The importance of this article is on the armed provisions for double-jointed WSNs.[16]

10.3.2   ENVIRONMENTAL SENSING

The time historical Protection Device Systems have improved to hood numerous solicitations of WSNs to ground technology research. This consists of sensing volcanoes, highlands, glaciers, forestry, and so on. Some of those are indexed underneath.[3]

10.3.3   AIR POLLUTION MONITORING

WSNs have been organized in numerous towns to screen the consideration of dangerous gases for motherlands.[9]

10.3.4   HEALTH

WSNs machines, billions of bucks of investment, and an international network of developers are riding a healthcare revolution. With a clear reoccurrence on backing and unnecessary average income conferring to customers, there are lots of healthcare WSN "killer apps" for subject monitoring, long-lasting disease management, and senior care.[7,9]

10.3.5   HOME

Standards-primarily based wireless sensor community adoption within the domestic has long been synonymous with domestic automation structures. Increasingly WSN will be deployed within those systems because the generation suits the equal large domestic utility areas that domestic automation has lengthy supported—energy control, domestic entertainment manipulate, healthcare and security, but they will additionally spearhead a more advert-hoc, consumer-led deployment and adoption of wireless automation inside homes.[9]

10.3.6   SPACE EXPLORATION

We can estimate the examination of the solar system with the reserve of ad hoc WSN, that is, networks wherein all nodes (either shifting or stationary) can both provide and relay statistics. It has been shown in Figure 10.1 how wireless sensors are used for various purposes. The issues of self-business commercial innovativeness and localization are the primary experiments to achievement over to achieve a consistent municipal for a diversity of assignments. We constituent out the kind of conservation and functioning forces that ought to face WSNs used for the universe investigation (Fig. 10.1).

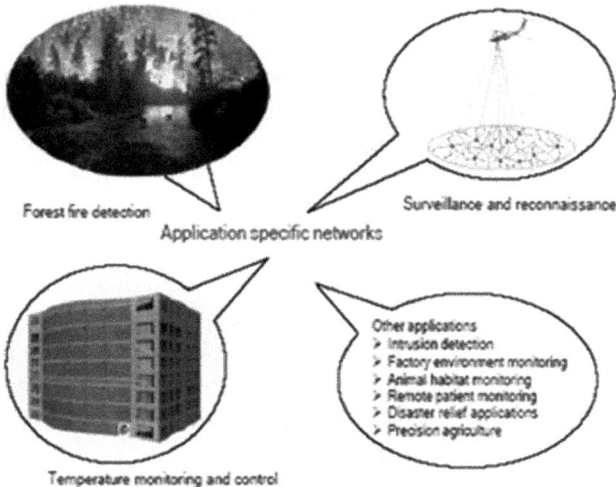

FIGURE 10.1   Application of wireless sensor network.

10.3.7   WATER MONITORING

Observing the superiority and smooth of liquid contains numerous goings-on such as examining the superiority of subversive or superficial liquid and safeguarding a country's water organization for the advantage of together social and animal. It may be used to safeguard the consumption of liquid.

10.3.8   NATURAL DISASTER PREVENTION

WSNs can be operative in precluding adversative significances of normal adversities, like floods. Wireless nodes have been organized positively in canals, where variations in liquid stages necessity be observed in actual period.

10.4   VIRTUALIZATION IN WIRELESS SENSOR NETWORKS

Virtualization is a warm topic substance in schmoozing in over-all; however, supreme of the virtualization approaches is considered for worried out networks. Wireless systems are exclusive in lots of significant issues. These methods want to be reformed to be concrete in wireless networks. Numerous wireless necessities are used to manage with all types of requirements. Though it correspondingly offerings a superior opportunity as virtualization might be the exact association to attach those necessities in a greenway. Additional expensive strength in wireless, networks is software well-defined transistor. It activities the competences of wireless communique and proposals an active instrument. It will add redecorate the inclusive concert of virtualization. Virtualization is set up via decoupling the request contributions from the fundamental physical network so that the usefulness contributions are not straight associated to their consistent unrestricted fundamentals, though are as an additional associated thru consistent/numerical associations clarify crossways the comprehensive municipal.. Thus, this emerging request has finished the idea of virtualization of these services a relatively useful generation for boosting resource sharing among unique stakeholder. WSNs are generally used to assemble info from the situation. The collective material is announced particularly to sinks or gateways that developed the endpoints wherein packages can repose and system such information. Though, plans strength also adopt from WSNs, an case-ambitious functioning perfect, so they may be informed on every instance rise a limited accurate environmental changes as conflicting to successively sympathetic the statistics providing infrequently.

10.5   INTERNET OF THINGS VS WIRELESS SENSOR NETWORKS

The different sensor and actuator nodes based on wireless networking knowledge are organized into the home atmosphere. These nodes produce the real-time data connected to the object practice and movement inside the home. All IoT solicitations requirement to consume one or supplementary strategies to gather the information afterward the situation. Sensors are necessary devices of horizontal materials. Unique of the extreme significant features of the IoT is circumstantial approachability, which is not probable without device knowledge. WSNs declarations to a gathering of specific devoted sensor with a communications organization. WSN is the underpinning of IoT applications. It is used for monitoring and recording the bodily situation conditions. IoT is the network of physical objects, and it is used to connect and exchange data over internet. In simple words, IoT is concept of things interaction with internet. IoT machine sensors immediately send their facts to the net.[12,16] It best simply contents the IoT programs necessities for lengthy-time period, low-charge and reliable provider, till refillable hardware and software program application systems are to be had, consisting of bendy Internet enabled servers to accumulate and method the sphere records for IoT applications. This paper supports the alteration for detectives confidential the WSNs substance can be summarizing as:

Comprehensive specifications for a stressful WSNs software program for a long-time period conservation checking that can be used to research the optimality of original WSNs answers.

Specs, layout apprehensions, and experimental consequences for stage components that form the characteristic IoT software program necessities of low price, excessive consistency, and extensive provision period.

A quick and conformation-loose discipline placement system appropriate for big measure IoT software deployments.[31]

10.6 APPLICATION REQUIREMENTS OF WSN FOR INTERNET OF THINGS

WSN is a bi-reversing wirelessly related system of instruments in a multi-hop smartness, produced from numerous bulges allotted in a sensor subject each connected to one or many sensors that would collect the item exceptional truths along with excessive temperature, moisture, bound, and so forth and then hop straight to the handing out tool. The sensing nodes conversation in multi-hop each sensor is a transceiver having an antenna, a micro-controller and an interfacing circuit for the sensors as a communication, actuation, and detecting unit correspondingly together with a source of strength which is probably each battery and any energy ingathering machinery (Fig. 10.2).

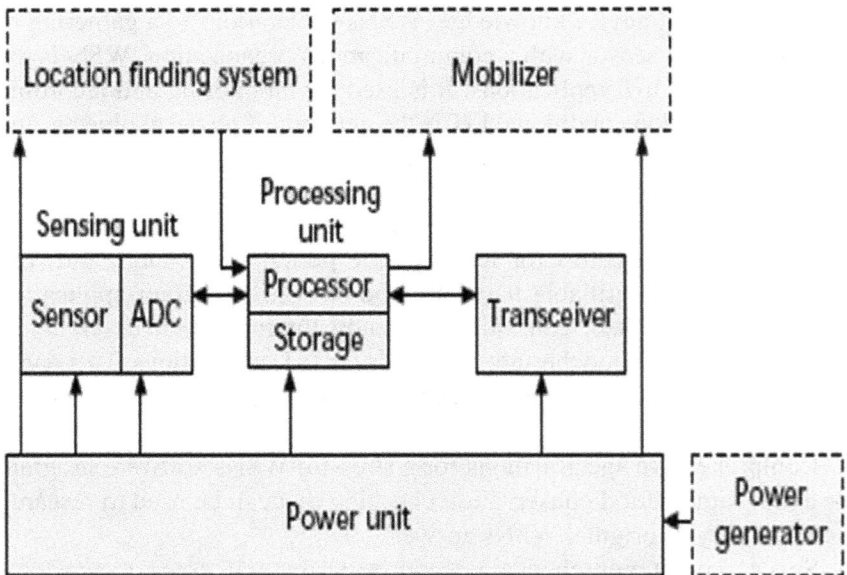

FIGURE 10.2   Components of sensor networks.

In this context, Wireless Sensor Network machinery is a necessary element of IoT because it contains of a gathering of sensor nodes associated finished wireless networks, accomplished of as long as numerical boundaries to the material-world things. Furthermore, as it happens in IoT, definite WSN applications can be functional to an extensive variation of areas like health, agriculture, logistics, wearable subtracting, and others. Though, current mechanism measured as WSNs solicitations are existence called as IoT solicitations without differentiating the novel topographies that describe this quantity. The application requirements between WSNs and IoT are exact comparable.

1. IoT is a new pattern that incorporates numerous machineries that previously occurred, such as WSN, RFID, Cloud Computing, Middleware systems, and end-user applications.
2. The number of technical records associated with WSNs has been deteriorating in current years, and this deteriorating is not due to WSNs is mislaying significance in nowadays, on the contrary, researchers are beginning to treat WSN as a technology integrated into the IoT ecosystem.
3. Although WSNs was originally conceived for local networks, WSNs applications may take advantage of Internet, even without being an essential requirement for WSNs.

10.7   CONCLUSION

In this chapter, we obtain the review of participating WSNs into the IoT in instruction to regulator electrical employments WSNs IoT, are not unmarried technology, however, alternatively represent complex systems the usage of numerous technologies from bodily communique layers to application programmers and are used in many utility areas and distinctive environments. This review examined the IoT allowed technology in rapports of knowing cities, dissimilar IoT, mist calculating, data mining, WSN-based facts centric IoT, cellular communique, background-consideration, virtualization, and real-time analytics. To comprehend the motivations of applying miscellaneous IoT components, we brought the necessities of different IoT factors with their time-honored goals. Next, we supplied a review of classes found out from dissimilar research that has been revised at some point of this paper. The IoT is progressive wherein IoT can be designated as an interconnection among recognizable strategies classified the internet assembly in identifying and observing methods. This chapter provides specific evaluation of WSNs. It also assesses the technology and characteristics of WSNs. Moreover, it affords evaluation of WSN programs and IoT applications. This study may be considered by investigators as an orientation opinion for those principally concerned in the study of this increasing field.

KEYWORDS

- **Internet of things**
- **wireless sensor networks**
- **distributed processing**
- **sensing**
- **communication**
- **localization**

REFERENCES

1. Li, L.; Xiaoguang, H.; Ke, C.; Ketai, H. The Applications of Wi-Fi Based Wireless Sensor Network in Internet of Things and Smart Grid. In *6th IEEE Conference on Industrial Electronics and Applications (ICIEA)*; 2011; pp 789–793.
2. Glombitza, N.; Pfisterer, D.; Fischer, S. Ltp: An Efficient Web Service Transport Protocol for Resource Constrained Devices. In *7th Annual IEEE Communications Society Conference on Sensor Mesh and Ad Hoc Communications and Networks (SECON)*; 2010; pp 1–9.
3. Mainetti, L.; Patrono, L.; Vilei, A. Evolution of Wireless Sensor Networks towards the Internet of Things: A Survey. In *19th International Conference on Software, Telecommunications and Computer Networks (SoftCOM)*; 2011; pp 1–6.
4. Yerra, R.; Bharathi, A.; Rajalakshmi, P.; Desai, U. Wsn Based Power Monitoring in Smart Grids. In *Seventh International Conference on Intelligent Sensors, Sensor Networks and Information Processing (ISSNIP)*; 2011; pp 401–406.
5. Perkins, C.; Royer, E. Ad-hoc on-demand Distance Vector routing. In *Second IEEE Workshop on Mobile Computing Systems and Applications. Proceedings.WMCSA '99*; 1999; pp 90–100.
6. Mikhaylov, K.; Tervonen, J. Evaluation of Power Efficiency for Digital Serial Interfaces of Microcontrollers. In *5th International Conference on New Technologies, Mobility and Security (NTMS)*; 2012; pp 1–5.
7. Akyildiz, I. F.; Su, W.; Sankarasubramaniam, Y.; Cayirci, E. A Survey on Sensor Networks. *IEEE Commun. Mag.* Aug **2002,** 102–114.
8. Ferrari, P.; Flammini, A.; Marioli, D.; Sisinni, E.; Taroni, A. Wired and Wi-Fi Sensor Networks for Business Packages. *Microelectr. J.* Sept **2009,** *40* (9), 1322–1336.
9. Ferrari, P.; Flammini, A.; Rizzi, M.; Sisinni, E. Improving Simulation of Wi-Fi Networked Manage Systems Based on Wireless HART. *Comput. Stand. Interf.* Nov **2013,** *35* (6), 605–615.

10. Dash, A. K.; Mohapatra, S.; Pattnaik, P. K. A Survey on Application of Wireless Sensor Community the Usage of Cloud Computing. *IJCSET* Dec **2010**, *1* (4), 50–55.
11. Durisic, M. P.; Tafa, Z.; Dimic, G.; Milutinovic, V. A Survey of Military Packages of Wireless Sensor Networks.*2012 Mediterranean Conference on Embedded Computing (MECO)*; 19–21 June 2012; pp 196, 199.
12. Depari, A.; Flammini, A.; Sisinni, E.; Vezzoli, A. A Wearable Smartphone-Based System for Electrocardiogram Acquisition. *2014 IEEE International Symposium on Medical Measurements and Applications Proceedings*; Lisbon, Portugal, 11–12 June 2014; pp 54–59.
13. Kaghyan, S.; Sarukhanyan, H. Accelerometer and GPS Sensor Mixture Primarily Based Gadget for Human Activity Recognition. *Computer Science and Information Technologies (CSIT)*; 2013, pp 1–9.
14. Zhang, Z.; Lv, T.; Su, X.; Gao, H. Dual Xor Inside the Air: A Community Coding Based Totally Retransmission Scheme for Wireless Broadcast Broadcasting. In *Communications (ICC), 2011 IEEE International Conference on*; IEEE, 2011; pp 1–6.
15. Al-Fuqaha, A.; Guizani, M.; Mohammadi, M.; Aledhari, M.; Ayyash, M. Internet of Things: A Survey on Allowing Technologies, Protocols, and Packages. *Communications Surveys & Tutorials, IEEE* **2015**, *17* (4), 2347–2376.
16. Shen, C.; Srisathapornphat, C.; Jaikaeo, C. Sensor Information Networking Architecture and Applications. *IEEE Pers. Common.*; Aug. **2001**; pp 52–59.
17. Shih et al., Physical Layer Driven Protocol and Algorithm Design for Energy-Efficient Wireless Sensor Networks. *Proc. ACM MobiCom '01*; Rome, Italy, July 2001; pp 272–286.
18. Wang, H.; Yip, L.; Maniezzo, D.; Chen, J.; Hudson, R.; Elson, J.; Yao, K. A Wi-Fi time Synchronized Cots Sensor Platform Component II–packages to Beamforming. In *Proceedings of IEEE CAS Workshop on Wireless Communications and Networking*; Pasadena, CA. 2002.
19. Wang, H.; Elson, J.; Girod, L.; Estrin, D.; Yao, K. Target Class and Localization in Habitat Monitoring. In *Proceedings of the IEEE ICASSP 2003*; Hong Kong, April 2003.
20. Wang, H.; Estrin, D.; Girod, L. Preprocessing in a Tiered Sensor Community for Habitat Monitoring.
21. Xu, Y.; Bien, S.; Mori, Y.; Heidemann, J.; Estrin, D. Topology Control Protocols to Conserve Energy in Wi-Fi Advert Hoc Networks. Technical Report 6, University of California, Los Angeles, Center for Embedded Networked Computing, January 2003. Submitted for booklet.
22. Heidemann, J.; Xu, Y.; Estrin, D.Geography-knowledgeable Energy Conservation for Advert Hoc Routing. In *Proceedings of the Seventh Annual ACM/IEEE International Conference on Mobile Computing and Networking (Mobicom 2001)*; Rome, Italy, July 2001.
23. Alkar, A. Z.; Buhur, U. An Internet Primarily Based Wi-Fi Home Automation Device for Multifunctional Devices. *Consumer Electronics, IEEE Transactions on* **2005**, *51*, 1169–1174.
24. Sharma, U.; Reddy, S. R. N. Design of Home/Office Automation Using Wireless Sensor Network. *Int. J. Comput. App.* **2012**, *43*, 53–60.

25. Liang, N.-S.; Fu, L.-C.; Wu, C.-L. An Incorporated, Flexible, and Internet-based Control Structure for Domestic Automation Machine in the Internet Generation. In *Robotics and Automation, 2002. Proceedings.ICRA'02.IEEE International Conference*; 2002; pp 1101–1106.

26. Rana, G.; Sharma, R. Emerging Human Resource Management Practices in Industry 4.0. *Strategic HR Rev.* **2019,** *18* (4), 176–181.

27. Feki, M. A.; Kawsar, F.; Boussard, M.; Trappeniers, L. The Internet of Things: The Next Technological Revolution. *Computer Feb* 2013, *46* (2), 24–25.

28. Cardone, G.; Cirri, A.; Corradi, A.; Foschini, L. The Participact Mobile Crowd Sensing Living Lab: The Tested for Smart Cities. *IEEE Commun. Mag.* Oct **2014,** *52* (10), 78–85.

29. Merentitis, A. et al. WSN Trends: Sensor Infrastructure Virtualization as a Driver towards the Evolution of the Internet of Things. *Proc. 7th Int. Conf. UBICOMM*; 2013; pp 113–118.

30. Abdelwahab, S.; Hamdaoui, B.; Guizani, M.; Rayes, A. Enabling Smart Cloud Services through Remote Sensing: An Internet of Everything Enabler. *IEEE Internet Things J.* June **2014,** 1 (3), 276–288.

31. Rajabzadeh, A.; Manashty, A. R.; Jahromi, Z. F. A Mobile Application for Smart House Remote Control System. *World Acad. Sci., Eng. Technol.* **2010,** 62, 80–86.

CHAPTER 11

Internet of Things Enabled Wireless Sensor Network

RAKESH KUMAR SAINI[1*], BHARTI SHARMA[1], GEETA RANA[2], and
RAVINDRA SHARMA[2]

[1]School of Computing, DIT University, Dehradun, Uttarakhand, India

[2]Swami Rami Himalayan University, Dehradun, Uttarakhand, India

*Corresponding author. E-mail: rakeshcool2008@gmail.com

ABSTRACT

The safety in slightly net beset the aspect of truth fullness, confidentiality, verification, anti-playback, and nonrepudiation. Safety has developed into the cutting edge of network management and application. Device nets are the nets of small devices that are expand aimlessly in numerous geological areas to gather advice around the weather. The core action of device lumps is to fold the statistics since the climate besides show the massed information to the improper position. Subsequently, these lumps increase in deserted weather; they are horizontal to the safety outbreaks. Safety is enhancing a major concern for Wireless Sensor Network (WSN) protocol designers because of the broad security-demanding applications of WSNs. It must be projecting that the method safety is focused on WSNS which needs a portion additional than what finds in other classes of network because WSN has its own uniqueness. One of the most alarming threats to WSN is the wormhole outbreaks, owing to their amount to employment direction-finding and solicitation statistics in actual period and aim dangerous compensations to the truthfulness, obtainability, and privacy of net statistics. In this chapter, we consider WSN security announce like major design challenging, security goals, threats, and attacks while collecting and processing data in WSNs.

11.1 INTRODUCTION

In the outdated announcement webs, the Exposed Organisms Interconnection covered construction has remained broadly implemented and has
assisted numerous infrastructures structures properly inside the previous;
though, growing sensor webs of these days are significantly tough in this
strategy viewpoint. The covered structure outlines a stack of protocol
coatings wherein each layer control inside its well-described feature and
borderline, and as a result permitting modifications to the underlying era
at every layer without implementing the want to trade the overall device
structure. This method has been a hit in its capability to offer modularity,
clearness, and standardization within the cord line networks, however,
strength be flawed within the Wi-Fi webs area.

Though wireless networks, such as moveable networks, Wi-Fi local
place networks (WLANs), mobile ad-hoc networks (MANETs) and wireless sensor networks (WSNs) are significantly one-of-a-kind in phrases of
their packages and architecture, a common subject in a majority of these
networks is the usage of the Wi-Fi channel for conversation. Unlike the
cord line networks, the Wi-Fi channel has numerous specific characteristics
that need to be taken into consideration while designing Wi-Fi networks.
WSNs can have a function to play for some of military purposes which
includes enemy movement detection and pressure monitoring. WSNs can
be used by the army for a number of purposes including tracking militant
interest in far flung areas and pressure protection. Being geared up with
appropriate sensors these networks can enable detection of enemy movement, identification of enemy force, and evaluation in their movement and
development. The attention of this text is on the military requirements
for bendy WSNs. Based on the main networking characteristics and army
use-instances, insight into particular army requirements is given to be able
to facilitate the reader's expertise of the operation of these networks within
the close to medium term (inside the next 3–8 years). Artificial Neural
Network detects and recognizes the activities of Daily life.[23] The article
frameworks the advancement of military sensor organizing devices by
recognizing three ages of sensors along with their capacities. Existing
developer solutions are offered and a top level view of some current
tailored merchandise for the army environment is given.[1,3] The time period
conservational network has advanced to cowl many packages of WSNs to
earth technological know-how studies. This comprises detecting volcanoes, mountains, glaciers, woods, and many others. Some of those are

recorded underneath. Area tracking is commonplace software of WSNs. In region watching, the Sensor Networks is arranged concluded a place in which a few phenomena are to be monitored.

First, the transmission environment of the wireless frequency necessitates decorative middle entrance regulator procedures for the station admittance and instant, the communicated indicator that broadcasts concluded the wireless middle is precious by reduction and upgrades additional quickly with remoteness as likened to the line stations.[4,5] A WSNs is a network that consists of scattered devices that are recycled to monitor the physical and conservational circumstances, such as temperature, sound, vibration, pressure, motion, and to supportively permit their information concluded the network to a main position. The more modern networks are bi-directional, likewise allowing regulators of device movement. The growth of WSNs was concerned by equipped solicitations such as battleground exploration; nowadays such nets are rummage-sale in numerous industrial and client solicitations, such as manufacturing method watching and control, mechanism fitness watching, etc.[6]

Device lumps container can be fictional as unimportant supercomputers, particularly straightforward in standings of their boundaries and their instruments. They typically contain of a dispensation component with imperfect computational control and incomplete remembrance, sensors is a broadcast expedient and an authority basis frequently in the form of a battery-operated. The two appearances of self-organization and localization are the main encounters to speechless to accomplish a trustworthy network for a diversity of undertakings.[7,8]

11.2 ISSUES IN WIRELESS SENSORS NETWORKS

The flexibility and convenience of WSNs come at a price. The multihop nature and the lack of fixed infrastructure add the complexities and design constraints in WSN. WSN has the following experiments:

- Incomplete wireless broadcast variety and channel capacity
- Transmission environment of the wireless standard
- Package wounded due to broadcast mistakes
- Flexibility-encouraged way variations
- Battery-operated restrictions
- Limited computational capabilities

- Limited storage
- Energy saving

11.3   COMPONENTS OF A SENSOR NODE

There are four basic components of sensor node, as shown in Figure 11.1: a sensing unit, a processing unit, a transceiver unit, and an electricity unit. They might likewise consume additional usefulness-founded mechanisms which comprise a part discovering scheme, vitality generator, and mobilizer. Sensing units are normally composed of subunits: sensors and analog-to-digital converters (ADCs). The analog alerts produced by way of the sensors primarily based on the determined phenomenon are converted to digital signals by using the ADC, after which nourished interested in the treating unit. The processing unit, which is characteristically related with a small storage unit, achieves the methods that make the sensor node cooperate with the conflicting nodes to achieve the allocated sensing duties.[9] A transceiver unit connects the node to the communal. One of the greatest important mechanisms of a sensor node is the energy unit. Power units can be maintained by consuming strong point rummaging unit's comprehensive of sun lockups. There are similarly other substitute components which can be application-dependent. Greatest of the device net direction-finding methods and identifying responsibilities necessitate information of position with great correctness.[10]

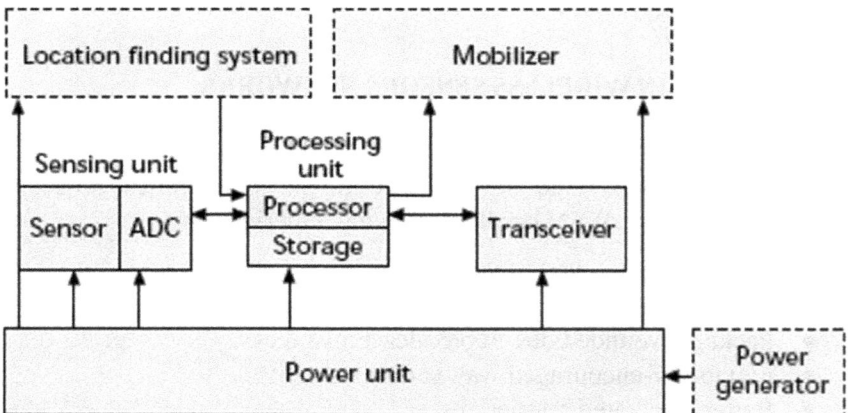

FIGURE 11.1   The components of a sensor node

11.4   WIRELESS SENSOR NETWORKS PROTOCOL STACK

The protocol stack used by the sink and sensor nodes as stated within the beneath in Figure 11.1 (The Sensor Network Protocol stack).[12] This conventions mountain associations strength and direction-finding appreciation, integrates records with interacting conventions, connects power efficaciously through the Wi-Fi middle, and encourages supportive energies of sensor nodes. The protocol stack consists of the physical layer, information hyperlink layer, network layer, transport layer, utility layer, electricity organization smooth, flexibility organization aircraft, and challenge control aircraft. The physical layer discourses the requirements of unpretentious though durable variation, broadcast, and getting plans. Figure 11.2 shows the sensor networks protocol stack.

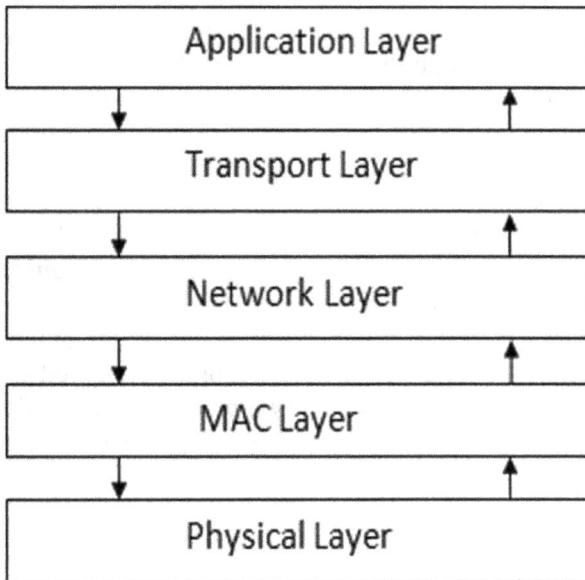

FIGURE 11.2   The sensor networks protocol stack.

Subsequently the surrounds is noisy and sensor nodes may be mobile, the medium get entry to controls protocol need to be strength-conscious and able to reduce impact with foreigners' pronounces. The network layer takings care of direction-finding the statistics provided through the shipping layer. The transportation coating facilitates to keep the movement of statistics if the WSNs utility calls for it. Conditional at the distinguishing obligations, unique varieties of solicitation software program can be constructed and used on the software layer. In accumulation, the control, movement, and assignment control planes reveal the electricity, motion, and undertaking distribution a number of the sensor nodes. These planes assist the sensor nodes manage the detecting enterprise and subordinate average control ingesting. The strength control smooth accomplishes exactly how a device knob uses its vitality.[13,24]

11.5 SECURITY REQUIREMENTS IN WSNS

In WSNs, sensor node is designed by different factor to achieve the assignment of detecting, treating, and communicating statistics. WSNs are charity for the numerous solicitations. The primary requirement of each one solicitation is to usage the protected net. Provided that safety to the device net is an exact confront circulate sideways through exchangeable its energy. Numerous safety hazards may change the working of these networks. Sensor network can also be operated unattended. Attackers may access the sensor nodes and extract all the confidential information from sensor nodes. They can also extract the data or can modify it. Attackers can also target the routing information which may misguide the traffic of the network. WSNs have some requirements for providing a secure communication. Overall safety necessities of WSNs are obtainability, privacy, truthfulness, and verification. Approximately other necessities recognized as unimportant necessities are source localization and data cleanliness. These necessities provide safety alongside outbreaks to the info communicated terminated the device net.[15]

11.6 SECURITY ISSUES IN WSN

The WSNs pattern is especially susceptible in contradiction of outside and inner occurrences. Consequently, it is essential to grow safety machineries and procedures to guard them. Better-quality safety is particularly signifi-cant for the accomplishment of the WSNs, since the information composed are frequently penetrating and the web is mainly susceptible. Although an amount of methods consume remained future to safety resolutions along-side numerous threats to the WSNs, greatest of which are founded on the covered strategy. Secure network is essential for cloud computing.[14] The objective of safety facilities in WSNs is to defend the info and incomes after outbreaks and misconduct. WSNs are second hand in many applications in military, eco-friendly, and fitness-associated ground.[16,17] These solicitations generally add the audit of delicate advice such as adversary act on the arena of crew in a building. Security is, therefore, crucial in WSNs. For the protected transmission of assorted types of info finished webs, a few cryptographic, steganography, and other approaches are used which are mine accepted.

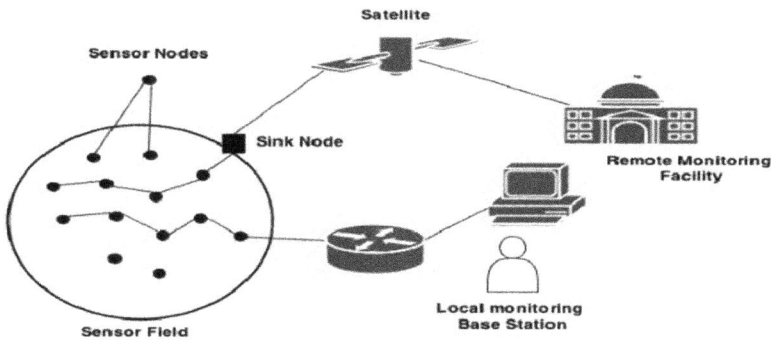

FIGURE 11.3   Sensor networks in distributed environment

In WSN, sensor nodes are deployed in distributed environment or in not reachable field (Fig. 11.3). These sensors are sense the particular environment and send data via satellite to base station or task manager node and then task manager node can take decision according to sense data. There are more security is required in distributed environment. There are some securities types that are require for secure communication in WSNs.

11.6.1   AVAILABILITY

Availability resolve even if a node has the capability to usage the assets and whether the net is accessible aimed at the communications to broadcast. A device net consumes to be prosperous across different safety attacks, and brunt must be downplay of a accomplish ambush. Though, it is acutely challenging to assure network opportunity interest to limited capability of entity device lumps to catch among hazard and breakdown.

11.6.2   SELF ORGANIZATION

Self-organization is one of the greatest significant features in a WSNs. Thousands of devices are organized in a conservation zone casually lacking allowing for the position issue. Nearby is not at all fixed organization available in WSN for the purpose of network management. Due to this feature of WSN, a great challenge of security is there. If self-organization is missing in an instrument net, the destruction subsequent from an occurrence or smooth the risky climate may be destructive.[18]

11.6.3   SECURE LOCALIZATION

Secure localization plays an energetic part in understanding the application context for WSNs. However, it is accessible to assorted hazard due to the open compromise area, the description of transmission radio, and ability pressure. Two kinds of attacks on localization process need to be checked. On the one hand, attackers may abduction, act like, clone nodes to betray the destination to a false area.[19]

11.6.4   CONFIDENTIALITY

Privacy is the capability to hide communications since an inactive enemy so that slightly communication connected by the device net remainders private. This is the greatest significant subject in net safety. WSNs sometime collect sensitive data required such as in military applications. Such deployed sensor node may require security aspects as data confidentiality. A sensor network should not leak sensor readings to its surrounding networks.

11.6.5   DATA AUTHENTICITY

Data authentication is critically aimed at numerous solicitations in wireless device webs. Since an attacker can easily implant messages, the headset requirements to create that information charity in any decision-making procedure arise since a credible basis. Commonly, the information confirmation permits an earpiece to authenticate that the information actually was referred by the assert correspondent.

11.6.6   DATA INTEGRITY

Information trustworthiness in device webs is wanted to promise the correctness of the archives and assign to the capability to verify that a communication consumes now not remained meddles through, changed though at the web. Even if the community has privacy adjust in region, nearby is nonetheless an action that the statistics' truthfulness consumes remained agree with the aid of adjustment.[19]

11.7   ATTACKS AND VULNERABILITIES IN WIRELESS SENSOR NETWORK

Different types of outbreaks are probable in WSNs. This safety outbreaks container be confidential conferring to dissimilar situation, such as the area of the enemies these security attacks in WSN should be on different protocol layer. Attacks can occur at any coating such as physical layer, data link layer, network layer, transport layer, and application layer.[17]

11.7.1   PHYSICAL LAYER ATTACKS

Wireless verbal exchange is the transmission via countryside. A conventional wireless indication is simple to jam or intercept. An Enemy ought to overhear or disrupt the provider of a Wi-Fi community physically. There are two sorts of assaults in Physical layer are:

(a) Jamming

Jamming attacks are a form of DoS assault wherein an opponent communicates a great-variety sign to disrupt communique. In the physical layer congestion assault, an opponent with an excessive transmission energy sign can jam the communique medium, as maximum WSN distributions activate on a solitary occurrence.

(b) Eavesdropping

The snooping assault is a critical security threat to a Wi-Fi sensor community because the snooping assault is a precondition for different attacks.

Conventional WSNs consist of wireless nodes geared up with omnidirectional antennas, which transmission wireless indicators in all instructions and are, therefore, disposed to the eavesdropping attacks.

11.7.2 DATALINK LAYER ATTACKS

The link layer is chargeable for multiplexing of data-streams, statistics body detection, medium get right of entry to control, and fault control. Attacks at this sediment encompass purposefully created smashes, useful reserve collapse, and wrongness in distribution.

(a) Collision

A collision attack is discovered on record hyperlink layers that knob neighbor-to-neighbor shipping together with road compromise. The full packet can be disturbed if an attacker is capable of generate blow of even a part of a transmission, CRC discrepancy and probable lack retransmission may be because of a single bit error.

(b) Exhaustion

Ingesting of a network's battery-operated energy can be satisfied via an exploration attack. A decide node ought to over and over ship, therefore, engrossing the battery energy more than required.[22]

11.7.3 NETWORK LAYER ATTACKS

The reason of this accretion is to make the records arrive from their ancestor to their vacation spot, even if each are not exactly related. Its intention is to discover the high-quality direction, making use of green routing algorithms. The forms of one-of-a-kind attacks located in this layer are:

(a) Wormhole

Wormhole assault is a grave attack wherein two attackers discover themselves strategically in the network. When the attacker nodes create a straight connection among every different within the network, then the wormhole enemy at one facet gets packets and transmissions them to the alternative facet of the network.

(b) Sinkhole

Sinkhole attack is such an attack that has been undermined hub that attempts to draw local area guests by methods for the market its phony steering supplant. One of the effects of sinkhole attack is that, it can be used to launch other attacks like selective forwarding attack, well known deceiving assault and drops or changed transmitting statistics.

(c) Hello Flood

Hello flood assault is the main assault in network layer. The Hello flood outbreaks can be as a result of a node that declares a Hello packet with very high power so that a big variety of nodes even a ways away inside the community pick out it because the discern node. All messages now want to be routed multihop to this parent, which will increase postpone. Hello messages are broadcast to a large number of nodes in a big region of the network.

11.7.4   TRANSPORT LAYER ATTACKS

The motive of shipping layer protocols in WSNs consists of putting in of give up-to-cease joining, quit to-quit dependable shipping of packets, drift manipulate, overcrowding manipulate, and clearance of stop-to-cease assembly. Outbreaks located on the delivery layer are:

(a) Flooding

A flooding outbreak happens in the network. It rapidly decreases the for the most part presentation of the WSNs.Desynchronization attacks

In desynchronization attack, an enemy interposes a lively joining among two lumps through communicating solid containers with spurious series statistics or manages streamers that desynchronize endpoints in order that they may retransmit the statistics. Header or full-packet authentication can defeat such an assault.

11.7.5   APPLICATION LAYER ATTACKS

Application layer DDoS assaults are considered to assault the software itself, that specialize in particular susceptibilities or matters, subsequent inside the solicitation now not being able to supply contented to the consumer. Application layer assaults can be noticed expending safety-targeted drift analysis; however, due to the fact that they are low-quantity attacks.[17]

11.8   ATTACKS AND VULNERABILITIES DETECTION TECHNIQUES IN WSNs

WSNs are defenseless alongside countless outbreaks. Consequently, we have to use approximately methods to safeguard information correctness, net functionality and its obtainability. As an outcome, we necessitate founding safety in WSNs with consideration to necessities and boundaries of these nets. Although the circumstance that the WSNs compromise a percentage, the safety encounters necessity be separated and undertaken consequently. Disappointment to organize this well-timed and adequately might concentrate it not relatively suitable to around the smallest objective similar any thoughtful of net. Intrusion Detection System is in charge of detecting, analyzing, and reporting unwanted intrusion that exploited the vulnerabilities of the networks and computer system. It acts as the second line of defense against attacks that preventive mechanism fails to address.[20,21]

11.9   CONCLUSIONS

Security has become a big concern for the data darkness as the various attacks are originating in WSNs. The arresting face of WSNs creates it very ask for to design strong security protocols while still maintaining low upward. WSNs have concerned considerable attention since together manufacturing and academe owed to its extensive series of solicitations such as atmosphere watching, showground attentiveness, medicinal healthcare, fighting investigation, and home-based employment organization. Therefore, info in the device net needs to be covered alongside numerous occurrences. Assailants might employ numerous security hazards making the WSN systems defenseless and changeable. In this chapter, we have tried to present most of the security attacks in WSNs with extensive study. In this chapter, we studied the challenges of WSNs and various attacks which found on the WSN protocol stack. This chapter mostly training the interruption discovery scheme in WSNs.

KEYWORDS

- **wireless sensor network**
- **safety**
- **occurrences**
- **passive and active attacks**
- **cryptographic attacks**

REFERENCES

1. Mouratidis, H.; Giorgini, P.; Manson, G. Using Security Attack scenarios to Analyse Security during Information System Design. In *The 6th International Conference on Enterprise Information Systems*, 2004.
2. Hunt, R.; *Network Security: The Principles of Threats, Attacks and Intrusions, Part1 and Part 2*; APRICOT, 2004.
3. Yick, J.; Mukherjee, B.; Ghosal, D. Wireless Sensor Network: A Survey. *Comput. Netw.* 2009, *52* (12), 2292–2330.
4. Dargie, W.; Poellabauer, C. *Fundamentals of Wireless Sensor Networks Theory and Practice*; John Wiley and Sons Ltd: Southern Gate, 2010.
5. Pathan, A. S. K.; Lee, H-W.; Hong, C. S. Security in Wireless Sensor Networks: Issues and Challenges. *Adv. Commun. Technol., IEEE Xplore*
6. Shanthi, S.; Rajan, E. G. Comprehensive Analysis of Security Attacks and Intrusion Detection System in Wireless Sensor Networks. *Next Generation Computing Technologies (NGCT)*, Dehradun, 2016, pp. 426-431, doi: 10.1109/NGCT.2016.7877454.
7. Bahl, N.; Sharma, A. K.; Verma, H. K. On Denial of Service Attacks for Wireless Sensor Networks. *Int. J. Comput. App.* Apr **2012**, *43* (6), 43–47.
8. Culler, D. E.; Hong, W. Wireless Sensor Networks. *Commun. ACM*, **2004**, *47* (6), 30–33.
9. Liang, Z.; Walters, J. P.; Chaudhary, V.; Shi, W. Wireless Sensor Network Security: A Survey. Security in Distributed Grid Mobile and Pervasive Computing, 2007. International Conference, IEEE Xplore
10. Bahl, N.; Sharma, A. K.; Verma, H. K. Impact of Physical Layer Jamming on Wireless Sensor Networks with Shadowing and Multicasting. *Int. J. Comput. Netw. Info. Security (IJCNIS), Hong Kong* June **2012**, *7*, 51–56. {ISSN: 2074-9090 (Print), ISSN: 2074-9104 (Online)}.
11. Chelli, K. Security Issues in Wireless Sensor Networks: Attacks and Countermeasures. *Proceedings of the World Congress on Engineering 2015*, Vol. 1; London, UK, 2015.
12. Jain, A.; Kant, K.; Tripathy, M. Security Solutions for Wireless Sensor Networks. In *Advanced Computing & Communication Technologies (ACCT), 2012 Second International Conference on IEEE*; 2012; pp 430–433.

13. Mishra, K. B.; Nikam, C. M.; Lakkadwala, P. Security Against Black Hole Attack in Wireless Sensor Network-a Review. In *Communication Systems and Network Technologies (CSNT), 2014 Fourth International Conference on IEEE*; 2014; pp 615–620.

14. Rana, G.; Sharma, R. Emerging Human Resource Management Practices in Industry 4.0. *Strategic HR Review* **2019**, *18* (4), 176–181.

15. Chelli P. Security Issues in Wireless Sensor Networks: Attacks and Countermeasures. *Proc. World Congr. Eng.* **2015**, *1*, WCE 2015, July 13 3, 2015; London, UK.

16. Dhar, M.; Singh, R. A Review of Security Issues and Denial of Service Attacks in Wireless Sensor Networks. *Int. J. Comput. Sci. Info. Technol. Res.* Mar **2015**, *3* (1); ISSN 2348–1196 (print), ISSN 2348-120X (online).

17. Pfleeger, C. P. S. L. *Security in Computing*, 3rd ed.; Prentice Hall 2003.

18. Chen, S.; Yang, G.; Chen, S. A Security Routing Mechanism against Sybil Attack for Wireless Sensor Networks. *International Conference on Communications and Mobile Computing (CMC)* **2010**, *1*, 142–146.

19. Yi, S.; Yongfeng, C.; LIANGRUI, T. A Multi-phase Key Pre-distribution Scheme based on Hash Chain. *9th International Conference on Fuzzy Systems and Knowledge Discovery (FSKD)*; 2012; pp 2061–2064.

20. Newsome, J.; SHI, E.; SONG, D.; PERRIG, A. The Sybil Attack in Sensor Networks: Analysis & Defences. In *Proceedings of the Third International Symposium on Information Processing in Sensor Networks*; ACM, 2004; pp 259–268.

21. Amuthavalli, K.; Bhuvaneswaran, L. Detection and Prevention of Sybil Attack in Wireless Sensor Network Employing Random Password Comparison Method. *J. Theor. Appl. Info. Technol.* **2014**, *67* (1), 236–246.

22. Blessey, P. M.; Princy, P. M. Defense Against the Sybil Attack with the Grid Based Transitory Master Key in Wireless Sensor Networks. *Int. J. Innov. Res. Comput. Commun. Eng.* **2015**, *3* (4), 3473–3480.

23. Jain, A.; Kant, K.; Tripathy, M. Security Solutions for Wireless Sensor Networks. In *Advanced Computing & Communication Technologies (ACCT), 2012 Second International Conference on IEEE*; 2012; pp 430–433.

24. Singh, R.; Anita, G.; Capoor, S.; Rana, G.; Sharma, R.; Agarwal, S. Internet of Things Enabled Robot Based Smart Room Automation and Localization System. In: *Internet of Things and Big Data Analytics for Smart Generation*; Balas, V., Solanki, V., Kumar, R., Khari, M., Eds.; Intelligent Systems Reference Library, vol 154; Springer: Cham, 2019.

Index

L

For Product Safety Concerns and Information please contact our EU
representative GPSR@taylorandfrancis.com
Taylor & Francis Verlag GmbH, Kaufingerstraße 24, 80331 München, Germany

* 9 7 8 1 7 7 4 6 3 9 2 1 4 *